Energy Efficiency in Shipping for Environmental Sustainability

This book examines the environmental impact of emissions from ships, setting out the mandatory regulations of MARPOL 73/78 and the Ship Energy Efficiency Management Plan, as well as new International Maritime Organization regulations such as MEPC 80.

The chapters provide an in-depth understanding of how ship staff can increase energy efficiency by reducing fuel consumption and using innovative technologies. Comprehensive coverage of energy audit methodology is offered to ensure compliance with energy efficiency requirements.

Written accessibly to suit a wide range of readers in the shipping industry, the book will be especially valuable for ship officers, trainees, and cadets.

Capt. Ashok Menon is from the 1977–78 cohort of Training Ship Rajendra. Since training, he has served in ascending ranks on ocean going vessels and was promoted to the rank of Captain in 1990. He served in this capacity for several years before turning to academia. He is currently the Head of the Department of Nautical Studies at FOSMA Maritime and Research organisation, Kolkata.

Energy Efficiency in Shipping for Environmental Sustainability

Ashok Menon

Routledge
Taylor & Francis Group

NEW YORK AND LONDON

Cover image: Snapshot freddy/Shutterstock.com

First published 2025
by Routledge
605 Third Avenue, New York, NY 10158

and by Routledge
4 Park Square, Milton Park, Abingdon, Oxon, OX14 4RN

Routledge is an imprint of the Taylor & Francis Group, an informa business

ISBN: 978-1-032-69877-9 (hbk)
ISBN: 978-1-032-70254-4 (pbk)
ISBN: 978-1-032-70256-8 (ebk)

DOI: 10.1201/9781032702568

Typeset in Sabon
by Apex CoVantage, LLC

Contents

Acknowledgements

I would like to thank my lovely wife Rajni, whose constant support and encouragement have been the wind beneath my sails. Her strength in holding down the fort while I was away sailing for months at a time, as well as when I was busy with my shore assignments, has been nothing short of extraordinary.

I'm also thankful to my brilliant daughter Shruti, whose keen insights and intellectual curiosity motivate me to become a better author.

No individual can flourish without the support of teachers, mentors and well-wishers, I am grateful to all of them.

Finally, I wish to thank my students, whose enthusiasm has always been a source of inspiration for me.

Chapter 1

Introduction

As energy is a scarce resource, depletion of energy sources would have a profound impact on our way of life. It is the sustenance of life, and we rely on it in various forms for our daily existence. We are dependent on it in various domains, such as transportation, agriculture, healthcare, and water supply. Thus, the importance of energy in our lives cannot be ignored, and it would not be wrong to say that if the sources of energy dry up, our way of life would be significantly impacted.

Energy, in its various forms, has been the lifeblood of maritime transportation for centuries, shaping the industry's evolution. Ships have an intimate connection with the environment, and constant effort is needed to ensure that they do not pollute it. Ships use fossil fuel as their source of energy, and the emission from ships exhaust contains large amounts of greenhouse gases (GHGs) which lead to global warming.

The greenhouse effect is, in fact, a natural phenomenon which traps the sun's energy and keeps the Earth's atmosphere within a range which can sustain life. However, the problem arises when the greenhouse effect is enhanced due to the presence of GHG, mainly carbon dioxide. The sun's rays have a short wavelength and can easily pierce the envelope of the atmosphere. But the heat energy reflected from the Earth is of long wavelength and cannot escape through the atmosphere, effectively trapping the sun's energy. The presence of carbon dioxide enhances the greenhouse effect and raises the temperature of the Earth. This is global warming which can have disastrous effects on our planet such as melting of glaciers, climate change, health risks and so on.

The only solution for global warming is to take preventive action by reducing the emission of GHGs. Since the emission of GHGs is linked to energy consumption across industries, the focus should be on reducing our dependence on fossil fuels by increasing energy efficiency and ultimately moving to alternate sources of energy.

In this era of increasing global awareness of environmental issues, energy efficiency in the maritime sector is a significant concern. It plays an important role in sustainable development by mitigating climate change, conserving resources, reducing operational costs, and improving air quality leading

DOI: 10.1201/9781032702568-1

to better health. From cutting down GHG emissions to alternative fuels, much can be done in this field. The industry faces a dual challenge: meeting the growing demand for global shipping while minimising its environmental impact. As the world increasingly recognises the urgency of preserving our oceans and curbing emissions, the maritime sector is putting into effect various innovations. In this book, we delve into the various means and methods of conserving energy and increasing energy efficiency in this sector. Although this is a global problem across all industries, this book will focus on the maritime industry and especially on shipboard energy efficiency. From the engine rooms of container ships to the decks of offshore platforms, this book looks at the strategies, technologies, and policies driving sustainability in the maritime realm. It explores how cleaner fuels, advanced propulsion systems, and eco-conscious practices can help us chart a new course towards a more environmentally friendly and economically viable maritime industry.

In this chapter, we will be discussing the importance of energy efficiency to reduce global warming and promoting energy sustainability. We shall also have a brief introduction to MARPOL and its various annexes.

Let's start with a brief history of the increasing importance of energy and its impact on the maritime industry today.

1.1 THE IMPORTANCE OF ENERGY EFFICIENCY

For a considerable part of our history, fossil fuels have played a dominant role in meeting the demand for energy. Coal, extracted since the 1800s, has been extensively used for heating and cooking to fuelling trains and ships. In the 1900s, oil emerged as an alternative for powering ships and other transport vehicles, gradually replacing coal. In 1872 George Brayten invented the first oil-fuelled internal combustion engine. Since then, commercial use of such engines has gone up by leaps and bounds, resulting in a huge uptick in oil consumption. Today, oil has become a primary fuel for transportation, while coal maintains its significance in electricity generation. Both coal and oil share a common origin: they are fossil fuels, formed over millions of years as organic materials were compressed and transformed beneath the Earth's surface. This renders them non-renewable resources, emphasising the need for conservation. Failure to do so could leave the world without its principal energy sources, which would be disastrous for all of us. Energy conservation is an urgent requirement on many fronts:

- Switching over completely to alternative sources of energy will take decades, until which time we have to manage with fossil fuels
- If we are profligate with fossil fuels, the next generation may run out of fossil fuels while not yet converted to alternative sources of energy
- Global warming associated with the use of fossil fuels that emit GHGs will one day make the Earth unbearable for habitation.

- Finally, the release of particulate matter, acid rains, and so on will not only affect vegetation but result in huge health issues for the population.

There is no doubt that sustainable use of our existing and limited sources of energy is extremely important, as is developing technologies and capabilities for the use of alternate sources of energy, including renewable energy. However, as we are all aware, although fossil fuels have been the primary source of energy in the maritime industry for a long time, they come with several significant disadvantages:

- The shipping sector is a notable source of CO_2 emissions, and these emissions can have a severe long-term impact on global warming and sea-level rise.
- The maritime industry's reliance on oil-based fuels also carries the risk of oil spills, which can lead to environmental disasters. Spills can harm marine ecosystems, wildlife, and coastal communities, causing long-lasting ecological and economic damage.
- Overreliance on these non-renewable resources in the maritime sector can lead to supply challenges and price volatility. As global demand for fossil fuels continues to rise, the maritime industry is vulnerable to these fluctuations.
- As the world shifts towards cleaner and more energy-efficient alternatives, the maritime industry's reliance on fossil fuels can lead to technological obsolescence and reduced competitiveness in the global market.

In response, there is a growing global push to transition to cleaner, more sustainable energy sources. However, this transition has its own problems, including intermittent supply, restricted scalability, and potentially unfavourable cost-benefit ratios. Hence, it becomes imperative to foster energy efficiency across various industries and applications worldwide. This approach is essential to ensure that we can continue to harness the advantages of energy consumption while dealing with the detrimental consequences it may entail.

1.2 ENERGY SUSTAINABILITY IN SHIPPING

Energy efficiency on ships not only helps reduce operational costs but also aligns with global efforts to combat climate change by decreasing the maritime industry's carbon footprint. As environmental regulations become stricter and fuel costs continue to rise, the adoption of energy-efficient technologies and practices in the shipping sector becomes increasingly important. Thus, the need of the hour is the sustainable use of energy. This means the responsible and efficient utilisation of energy resources in a manner that meets current needs without compromising the ability of future generations to meet their own energy needs. The following are the steps required to conserve energy and reduce its harmful effects:

- **Energy Efficiency on Ships:** Improving energy efficiency is a key component of efforts to reduce the emission of GHGs from ships. The International Maritime Organization (IMO) has made the Ship Energy Efficiency Management Plan (SEEMP) mandatory for all ships. This plan focuses on various measures to reduce the fuel consumption and thereby increase the energy efficiency. Embracing energy-efficient practices like optimised route planning, slow steaming, and hull and propeller cleaning can reduce fuel consumption and emissions. Regular maintenance and retrofitting with energy-saving technologies, such as air lubrication systems, can also improve efficiency.
- **Alternative Fuels and Propulsion:** Moving away from fossil fuels to cleaner alternatives like hydrogen, ammonia, methanol, and biofuels can significantly reduce GHG emissions and air pollutants. Implementing innovative propulsion technologies, such as electric and hybrid systems, can also enhance energy efficiency.
- **Regulatory Compliance:** Adherence to international maritime regulations, such as the IMO's regulations on emissions, ballast water management, and fuel quality, is essential for reducing the environmental impact of shipping. Compliance with emission control areas (ECAs) and the Energy Efficiency Existing Ship Index (EEXI) is also critical.
- **Waste Management:** Proper waste disposal is essential to prevent pollution of the seas. Implementation of waste management plans and adherence to MARPOL Annexes IV and V are crucial.
- **Innovative Design and Retrofitting:** New ship designs that prioritise sustainability and energy efficiency, such as Energy Efficiency Design Index (EEDI)-compliant vessels, are crucial. Retrofitting older vessels with energy-efficient technologies can also enhance sustainability.
- **Crew Training and Awareness:** Training crew members in best practices for sustainability, safety, and environmental protection is vital. Well-informed and proactive crews can play a significant role in minimising the industry's environmental footprint.
- **Collaboration and Partnerships:** Collaboration between governments, industry stakeholders, and organisations can drive collective efforts to achieve sustainability in the shipping sector. Partnerships for research, sharing best practices, and addressing common challenges are instrumental in this regard.

We will discuss these points in greater detail later in this book. Energy sustainability is important to reduce the environmental impact of the maritime industry, which traditionally relies on fossil fuels. A coordinated focus on energy sustainability not only prevents environmental harm but also ensures the long-term viability and resilience of the maritime sector. It is evident that the sustainable use of energy is essential not only for environmental preservation but also for long-term economic stability and social well-being. By adopting a holistic approach to energy management, we can create a more resilient and sustainable energy future for ourselves and future generations.

Let's look at a couple of organisations that have been foremost in their efforts to highlight the importance of implementing energy efficiency to reduce energy consumption and prevent global warming.

1.3 ORGANISATIONS IMPLEMENTING ENERGY EFFICIENCY

Organisations like the International Energy Agency (IEA) and the International Maritime Organization work towards mitigating climate change by reducing GHG emissions, enhance energy security, and conserve resources.

1.3.1 The International Energy Agency

The IEA is an intergovernmental agency tasked with promoting environmental sustainability by cooperation and collaboration among member countries, non-member countries, and other stakeholders. It consists of 30 members, mainly from the Organization for Economic Co-operation and Development (OECD).

The IEA has issued the World Energy Outlook 2022,[1] which sets forth the roadmap for net zero emissions and the key energy trends. The study put the spot light on the global energy crisis and current energy trends. According to this report, the only way to achieve the goal of net zero emissions by 2050 is to scale up the production of clean energy and cut down on fossil fuels. The good news in the study is that the use of renewable energy is increasing at a faster rate than other sources of energy, with wind and solar power having dominance over others. Overall, this in-depth study gives a good look at past trends, as well as future projections related to energy consumption and production. The study found that energy prices have been steadily increasing over the years, and high fuel prices resulted in higher electricity generation costs. The study also found that this will probably lead to a cost advantage for clean energy technologies. CO_2 emissions from the use of fossil fuels, which was 36.6 Gt in 2021, would plateau and then fall to 32 Gt in 2050. A smooth transition to clean energy will require major investments in technology and infrastructure.

1.3.2 The International Maritime Organization

The IMO[2] is an agency of the United Nations and works with stakeholders for safe and sustainable shipping. The key objectives of the IMO are ensuring the safety and security of shipping, prevention of pollution, efficient operations, and promoting sustainable maritime operations. To achieve these objectives, IMO works with member states and maintains a regulatory framework, including conventions, codes, resolutions, and guidelines. These requirements are enforced by flag states (for ships registered under their flag) and port state control (for foreign ships visiting the port). The most important conventions are the International Convention for the Safety of Life at Sea 1974, as amended (SOLAS), and the International Convention for the Prevention of Pollution from Ships 1973, as amended by the protocol of 1978 (MARPOL 73/78).

IMO promulgates their conventions and regulations which are mandatorily required to be followed on board the ships. They adopt various resolutions at meetings of member countries dealing with various aspects of safety, security, and environmental protection.

Energy sustainability is important to reduce the environmental impact of the maritime industry, which traditionally relies on fossil fuels. A coordinated focus on energy sustainability not only prevents environmental harm but also ensures the long-term viability and resilience of the maritime sector. Recent amendments to MARPOL have put the focus on air pollution and reduction of GHGs, leading to a sustainable shipping industry.

1.4 INTRODUCING MARPOL

Energy efficiency in the maritime industry is important due to the significant energy consumption associated with ships. Improving energy efficiency in ships is essential for reducing fuel consumption, cutting GHG emissions, and minimising operating costs.

The International Convention for the Prevention of Pollution from Ships (MARPOL) 1973, as amended by the protocol of 1978 (MARPOL 73/78), is the IMO publication that deals with ways and means to curb pollution from ships. MARPOL 73 was the response to the grounding of the tanker *Torrey Canyon* in March 1967, off the southwest coast of England, which resulted in a massive oil spill of more than 1,00,000 tons. The Marine Environment Protection Committee (MEPC) was entrusted with drafting a convention to address various forms of pollution from ships. However, before MARPOL 73 could enter into force, a series of incidents involving oil spills took place, culminating in the wrecking of *Amoco Cadiz* in March 1978. This resulted in a major oil spill of more than 2,20,000 tons off the coast of Brittany in France. Lessons learnt from these incidents were incorporated into MARPOL and the protocol of 1978. One of the biggest was the South Korean tanker Sea Star, which collided with the Brazilian-registered Hosta Barbosa, resulting in an oil spill of nearly 115,000 tons in December 1972. These incidents spurred IMO to learn from past mistakes and include some new requirements in MARPOL, which was the protocol of 1978.

MARPOL 73/78 entered into force on 2 October 1983 in a phased manner. All vessels trading internationally must comply with the regulations of MARPOL. Flag administrations, that is, the country where the ships are registered, will ensure that the ship is complying with all mandatory requirements including MARPOL. Port state controls of various countries visit foreign ships in their water and confirm compliance with these requirements. If ships are found to be not complying, then they may even be detained and prevented from sailing out till corrective actions are taken.

Now, let's take a brief look at the various annexes of MARPOL 73/78[3] and their efforts to protect the environment:

- **Annex I:** This annex of MARPOL deals with the discharge of oil into the sea from tank washings and machinery spaces. It lays down stringent criteria for such discharge from both the cargo spaces and machinery spaces. Discharge from cargo spaces is permitted only when the cargo tanks are being washed, subject to several criteria. Discharge from machinery spaces is permitted when the vessel is fitted with a 15 ppm alarm and auto-stop system (for vessels greater than 10,000 GT). Vessels must maintain an oil record book to record all activities related to the oil cargoes or from machinery spaces. An approved Shipboard Oil Pollution Emergency Plan (SOPEP) should be available on board and the requirements of the plan are to be complied with. Oil tankers must have an oil discharge monitoring system (ODMCS) to record the discharge of oil mixtures from cargo spaces. All vessels must be equipped with an oily water separator (OWS) and oil filtering equipment to separate the oil from bilge water before discharging it.
- **Annex II:** This annex deals with the discharge of residues into the sea from vessels carrying noxious liquids in bulk. It divides the noxious liquid substances into three categories, namely X, Y, and Z. Vessels carrying certain chemical cargoes that pose a major threat to the environment (category X) must conduct tank washing after discharge and before departure from port. The residues from the washing will have to be landed ashore in a reception facility. Vessels must maintain a cargo record book to record all cargo operations, tank cleaning, and other activities related to the carriage of noxious liquid substances in bulk.
- **Annex III:** This deals with the carriage of dangerous substances in packaged form. Vessels must follow the International Maritime Dangerous Goods Code when carrying such types of cargo. This code details the segregation and separation of various classes of dangerous goods as well as the actions to take when these substances are spilt or on fire. Contact with these dangerous substances can lead to burns and other adverse health effects, and the emergency schedule gives the actions to take in such cases.
- **Annex IV:** This deals with the discharge of ship-generated sewage and requires vessels to be fitted with an approved sewage treatment plant.

If such equipment is not fitted on board, then the vessel must collect the sewage in a holding tank and discharge it 12 nautical miles from the nearest land.

- **Annex V:** Ships generate garbage daily. Annex V states the criteria for discharging garbage at sea. As per the criteria, only food waste can be disposed of at sea and that too subject to certain conditions. Other general garbage must be landed at reception facilities ashore, and the landing receipts maintained on board as evidence to be shown to the authorities as required.

- **Annex VI:** This is the most relevant annex for air pollution. It essentially lays down requirements for ships' exhaust gas emissions such as sulphur oxides, nitrogen oxides, carbon dioxide, particulate matter, and so on. In addition, it covers other forms of air pollution from ships, such as ozone-depleting substances and pollution from incinerators.

  The annex sets limits on sulphur content in fuel oil, establishes standards for NO_x emissions, and designates ECAs with stricter requirements. It also promotes energy efficiency through the EEDI for new ships and requires SEEMP. Compliance with Annex VI is essential for reducing the environmental impact of maritime transportation and promoting sustainable shipping.

  We shall discuss Annex VI in later chapters.

MARPOL is the most important instrument for IMO to achieve their target of reduction of GHGs. It promotes sustainable shipping practices by tackling environmental issues, encourages efficiency enhancements, promotes innovation, safeguards human health, and strengthens regulatory structures. Compliance with MARPOL regulations is essential for achieving sustainable shipping.

Before we move on, let's look at some other conventions that have been implemented to help protect the environment.

1.5 OTHER CONVENTIONS RELATED TO ENVIRONMENTAL POLLUTION

MARPOL is the primary convention dealing with the prevention of environmental pollution. However, there are other conventions that also relate to the protection of the environment. From control of ballast water discharge to oil pollution liabilities, these conventions cover most of the aspects of environmental damage from ship-related activities. Some of these conventions are as follows:

- **International Convention on Civil Liability for Oil Pollution Damage:** The Civil Liability Convention (CLC) was brought into force to deal with cases of pollution when the pollution clean-up costs and claims can be covered by insurance. In the case of Torry Canyon, the ship-owner got away with paying only a fraction of the clean-up costs and related claims from fishermen and the tourism industry. Thus, IMO came up with the CLC, which makes it mandatory for all tankers to obtain insurance for their maximum liability as per the convention. The CLC imposes strict but limited liability on the ship-owners for oil pollution. It is strict because the ship-owners must obtain insurance coverage for their maximum liability. Based on this, the flag administration of the vessel will issue the CLC certificate. It is limited because the conventions impose a limit on the maximum liability of the ship-owner. However, if there is wilful negligence on the part of the ship-owner, then they will be exposed to unlimited liability. The CLC was augmented by the fund convention, which makes the oil importers of different countries contribute to the fund maintained by IOPC London. When an oil spill occurs and the expenses are over and above the CLC limits, then the fund steps in to defray the expenses up to a certain limit.

- **The International Convention for the Control and Management of Ships' Ballast Water and Sediments, 2004 (BWM Convention):** The convention entered into force on 8 September 2017. It has long been known that ballast water was the source for harmful aquatic organisms to travel across the globe. Thus, the convention made it mandatory for new ships to be fitted with a ballast water treatment (BWT) system meeting the D-2 standard. According to this, existing ships must retrofit the BWT system during their first International Oil Pollution Prevention Certificate after 7 September 2019. This means that by 8 September 2024, all applicable vessels will be fitted with the BWT system. Till date, the ship will have to conduct an exchange of ballast water more than 200 nm from the nearest land in order to flush out the aquatic organisms and have clean deep-sea ballast in their ballast tanks. The ballast exchange has to be carried out as per the D-1 standard for those ships not fitted with a BWT plant. Such ships will have to exchange their ballast water in open seas, 200 miles from the nearest land and in waters 200 metres deep. D-2 is the performance standard for the treated ballast water from the BWT plant for ships fitted with one. It specifies the maximum number of organisms that are permitted to be discharged in the treated ballast water.
- **International Convention on Salvage (Salvage Convention):** Salvage is a voluntary service to save a ship in distress. Salvage serves two purposes, mainly saving property and the environment. If a ship that has run aground is on fire or in any form of distress, the salvors will make every attempt to save the ship, thus minimising or preventing losses for the ship-owner as well as the cargo owners. Furthermore, if the ship had become a wreck or capsized, substantial oil pollution would have occurred, resulting in environmental degradation. To ensure smooth

and efficient salvage and streamline this process, the salvage convention was adopted by the IMO. The Salvage Convention provides a framework for cooperation and compensation in the event of maritime salvage operations, promoting environmental protection and preventing pollution from shipwrecks. Salvaging a ship instead of letting it remain a hazard to shipping and the environment promotes sustainability by recovering resources, reducing waste, preventing pollution, protecting habitats, preventing accidents, and minimising environmental impact.

- **Nairobi International Convention on the Removal of Wrecks:** This convention deals with vessels that have become wrecked. Very often, ship-owners desert wrecked vessels, leaving the coastal state to remove them at their own cost. A vessel that is wrecked will not only be a hazard to navigation but also pollute the environment. This convention places the responsibility for removing hazardous wrecks on the ship-owners. The ship-owners must obtain insurance for the wreck removal costs. Accordingly, when the ship is wrecked, this insurance coverage can be enforced to cover expenses for the wreck removal. The amount of the insurance coverage required depends on the gross tonnage of the ship. The insurance cover may be obtained from any approved insurer, based on which the administration issues the Nairobi Wreck Removal Certificate.

MARPOL and other related conventions comprehensively cover the various aspects of environmental pollution from ships. Essentially, it is Annex VI that deals in detail with air pollution and energy efficiency, and we shall discuss these regulations in detail along with upcoming amendments.

In addition to these conventions, there are other initiatives that are discussed in later chapters, such as technological innovations, industry initiatives, collaboration between stakeholders, green ports, and so on. Collectively, these efforts promote sustainable practices within the maritime industry, with the objective of energy efficiency, environmental protection, and long-term viability.

1.6 SUMMARY

In this chapter, we have seen the chain of events leading to climate change. An understanding of this sequence is important if we are to delve into the deeper subject of energy efficiency and sustainability. The use of fossil fuels leads to GHG emissions, which ultimately lead to global warming and climate change. We have briefly seen the ways and means to conserve energy. We introduced MARPOL along with its various annexes. Further, we have discussed the various conventions regarding environmental protection. This should give you a fair idea of the importance of energy efficiency and environmental protection.

In the next chapter, we will discuss why energy-efficient operations are so important on-board ships, and how we can start implementing them.

BIBLIOGRAPHY

1. International Energy Association. *World Energy Outlook 2022.* https://iea. blob.core.windows.net/assets/830fe099-5530-48f2-a7c1-11f35d510983/ WorldEnergyOutlook2022.pdf.

    Details about the World Energy Outlook and the roadmap to net zero emissions along with the outlook for energy demand and electricity.
2. International Maritime Organization. (2024). www.imo.org.

    The IMO website is the central source for information on IMO conventions, publications, guidelines, events, and regulations to enhance maritime activities and safe operations.
3. International Maritime Organization (IMO). *International Convention for the Prevention of Pollution from Ships (MARPOL).* https://www.imo.org/ en/about/Conventions/Pages/International-Convention-for-the-Prevention-of-Pollution-from-Ships-(MARPOL).aspx.

    MARPOL is the international convention covering various aspects of pollution from oil, chemicals, hazardous substances, sewage, garbage as well as emissions from ships.

Chapter 2

Energy Efficient Ship Operations

Most of the world's transport depends on fossil fuels to provide the required energy. Fossil fuels by their very nature have a limited source of supply. Thus, it is imperative on the part of policymakers to ensure energy efficiency so that available energy is conserved and not wasted through systematic inefficiencies. Ships' engines burn fossil fuels to generate energy, which results in exhaust gases being released into the atmosphere. These gases contain not only carbon dioxide, which is a major greenhouse gas (GHG), but also other by-products of combustion, such as methane and nitrous oxide (these can also be considered as GHGs but to a lesser extent).

According to the International Maritime Organization (IMO), the shipping industry accounts for almost 3% of the total GHG emissions globally. This is not a small figure. While the global industry has been sitting up and taking notice of the urgent need to reduce GHG emissions, the shipping industry has been recalcitrant in taking a giant step forward in this field for too long. It is only now that the IMO has brought in a slew of measures to address this urgent problem.

However, the industry still faces several challenges in implementing these measures. As with any new regulations, there will be resistance to change on part of the ship-owners, managers and even ship staff. Moreover, there will be an additional cost involved, such as incorporating new technology, additional training needs, and so on.

In this chapter, we will discuss the current state of sustainability, highlighting the countries that are the top emitters of carbon dioxide and the share of the shipping industry in worldwide emissions. We will also discuss the solutions that are currently available to reduce the fuel consumption on ships, resulting in cutting down on their GHG emissions.

2.1 CURRENT STATE OF SUSTAINABILITY

The shipping industry is increasingly focused on sustainability because of regulatory requirements, awareness of environmental concerns, and market demands. But much work needs to be done if the shipping industry is to meet the United Nations goals of sustainable development.

DOI: 10.1201/9781032702568-2

It is a fact that even today most ships being delivered are without the technological advances available currently. This is because typically, ship-owners are unwilling to invest in technology. They prefer to comply with the bare minimum regulations, which will ensure that they do not run afoul of the enforcing agencies, such as port state and flag state controls, and can continue trading as usual.

According to the IEA, the total carbon dioxide emissions in 2021 were 36.3 billion metric tons, an increase of 6% over the previous year. This figure is set to increase annually if urgent measures are not taken to control the emission. China accounted for almost 30% and the United States for almost 14% of the global emissions. Thus, the major efforts must come from these countries to reduce their emissions and make the world free from the threat of continued global warming. Table 2.1 shows the carbon dioxide emissions worldwide in the years 2010 and 2021. China's emissions have increased by over 33%, while US emissions have reduced by 12% during the period.

The share of the shipping industry in the global carbon dioxide emissions from fossil fuels is nearly 3%. Table 2.2 shows the statistics as per studies conducted by the IMO.

As per information published in Lloyd's list, carbon dioxide emissions in the shipping industry show a year-on-year rise of 4.9% in 2021. This is a big setback for the shipping industry's continuous efforts to reduce GHG emissions. It is evident that the shipping industry is responsible for a significant portion of the global GHG emissions, and this figure is increasing constantly. As per studies conducted by the European Commission, if immediate action

Table 2.1 Top CO_2 emitters in 2010 and 2021 (in million metric tons)

Country	2010	2021
China	8,617	11,472
USA	5,681	5,007
India	1,676	2,710
Russia	1,626	1,756
Japan	1,215	1,067

Source: Statista, 2023[4]

Table 2.2 Global CO_2 emissions from shipping (in million metric tonnes)

Year	Global anthropogenic CO_2 emissions	CO_2 emissions from shipping	Shipping emissions as a percentage of global emissions
2012	34,793	962	2.76%
2013	34,959	957	2.74%
2014	35,225	964	2.74%
2015	35,239	991	2.81%
2016	35,380	1,026	2.90%
2017	35,810	1,064	2.97%
2018	36,573	1,056	2.89%

Source: IMO, 2020[5]

is not taken, emissions from shipping are likely to increase by up to 130% by 2050 as compared to 2008 levels. Hence there is a need for urgent action to control and reduce GHG emissions from ships. This includes continued collaboration between industry stakeholders, governments, and international organisations to overcome challenges and drive innovation towards cleaner and more efficient maritime transportation.

Before we look at how these issues can be addressed, let's review the impacts of shipboard emissions.

2.2 SHIPBOARD EMISSIONS

Although we are primarily concerned with air pollution and GHG emissions, let us have a look at the various ways in which the operation of ships can affect the environment. Ships' main engines use fuel for energy that is used by the engines to provide the thrust to move the ship forward. The auxiliary engines use fuel to provide electricity for lighting and other auxiliary equipment on board. The fuels used by the ships are primarily fossil fuels, and thus their exhaust gases contain a large number of harmful substances, which are as follows:

- **Carbon Dioxide:** Carbon dioxide (CO_2) is emitted from ships due to the combustion of fossil fuels in ship engines. It is a GHG and leads to global warming by trapping the sun's heat. Further carbon dioxide in the atmosphere absorbs the radiation of the sun and re-emits it, causing further warming of the atmosphere. Global warming leads to climatic changes, melting of ice caps, increase in sea level, increase in the seawater temperature affecting marine flora and fauna, and so on. Carbon dioxide in the atmosphere also has harmful effects on humans, animals, and the environment in general. Carbon dioxide can be absorbed by seawater, increasing the acidity of the oceans and leading to adverse effects on marine life and the ecosystem.
- **Nitrogen Oxides:** Nitrogen is an inert gas and does not usually combust. However, in a combustion chamber, nitrogen can react to form nitrogen oxide (NO_x) because of the high temperature and pressure. Nitrogen oxides are harmful gases that act in many ways. First, they can be directly deposited on plants and may prove to be toxic to them. Second, if present in the atmosphere, they interact with rainwater, causing acid rain, which damages both the plant life and the soil. This acid rain can also have a detrimental effect on the oceans, thereby affecting marine life in general. Third, when inhaled, even very small amounts can cause severe health problems. Finally, they can interact with volatile organic compounds (VOCs), resulting in ground-level ozone, which has a detrimental effect on human health, causing respiratory distress, allergies, cardiovascular problems, decreases in crop productivity, and so on.
- **Sulphur Oxides:** Sulphur oxides are produced in the ship's engines during the combustion of fossil fuels due to the presence of sulphur in

fossil fuels. These are also harmful gases that affect the atmosphere. They can cause acid rain, affecting plant life, soil pollution, and so on. They also increase the acidity of oceans, affecting marine life. They directly affect humans as even a negligible quantity can cause severe respiratory damage and distress, eye irritation, and other health issues. They can also cause corrosion of metal structures as the sulphur oxides can form sulphuric acid in the presence of moisture.

- **Particulate Matter:** This refers to minute particles of carbon and other pollutants released in exhaust gases as a result of combustion. When particulate matter is released into the atmosphere, it may travel to the respiratory tract, causing deep distress in the form of coughing, sneezing, and other diseases.

Thus, ships' exhaust gases cause several problems for the environment in general and humans in particular. There is no doubt that shipping is the most energy-efficient form of transport. But the air pollution caused by ships, if not controlled, can cause severe environmental problems. As per UN statistics, the Earth is 1.1° warmer now than it was in the 1800s. To combat the menace of air pollution, the United Nations has introduced the concept of net zero. This means cutting down on GHG emissions across all fronts to contain global warming to an increase of 1.5°.

2.3 WHAT IS THE SOLUTION?

We are aware that, just as in other sectors, global warming is the main issue facing the maritime industry today, primarily due to the release of GHGs. The obvious solution would be to reduce the consumption of fossil fuels and, ultimately, transition to alternative, eco-friendly fuels. However, this is easier said than done, as this transition is complex, spanning from the practical challenges of implementation to the various cost-benefit considerations. These complexities underscore the magnitude of the transformation required to steer the maritime industry towards a sustainable and environmentally responsible future.

To address these challenges, MARPOL Annex VI has imposed a sulphur cap on marine fuels as follows:

- From 1 January 2015, the sulphur content in marine fuels within ECA areas is not to exceed 0.1% m/m.
- From 1 January 2020, the sulphur content in marine fuels outside ECA areas is not to exceed 0.5% m/m.

For the emission of NO_x, Annex VI has the following limitations:

- For nitrogen oxides (NO_x) emission, Marine Engines installed after 1 January 2000 will have to comply with the "Tier I" emission limits, that is, not more than 17.0 g/kWh for engines with rated rpm less than 130.

- For NO_x emission, marine engines installed after 1 January 2011 will have to comply with the "Tier II" emission limits, that is, not more than 14.4 g/kWh for engines with rated rpm less than 130.
- For NO_x emission, marine engines installed after 1 January 2016 will have to comply with the "Tier III" emission limits that is, not more than 3.4 g/kWh for engines with rated rpm less than 130. Tier III limits apply to ECA areas.

Now, let's briefly discuss some measures that can be taken on board ships to contribute to energy efficiency.

2.3.1 Reducing Fuel Consumption

The emission of CO_2 and other noxious gases has a direct correlation with fuel consumption. It is a known fact that fuel-efficient engines are the best way to lower air pollution. The ship's main engine burns heavy fuel oil, leading to a high level of air pollution in the standard two-stroke marine engine. Thus, increasing fuel efficiency is the focus of IMO's initiatives in this field. The move to reduce the fuel consumption has many benefits:

- Scarce fossil fuels are not unduly wasted and are preserved for future generations. This will ensure that fossil fuels will not disappear altogether within a few years but are instead conserved for a longer period of time, during which period alternative fuels can be introduced.
- Cost-benefit for the ship-owners. Fuel is one of the major operational costs for the ship, and thus, reduced fuel consumption translates into cost savings for ship-owners.
- A reduction in fuel consumption will directly reduce the galloping demand for fossil fuels. This will result in an arrest in the increase of fuel prices, providing a double relief for the ship-owners as well as the customers in general as freight rates will come down over a period.
- One of the most important effects of reduced fuel consumption is a reduction in the emission of noxious substances such as sulphur dioxides, NO_x, and carbon dioxide. GHG emissions have a negative impact on the environment, and any reduction in their emissions will be highly beneficial to society as a whole by improving the air quality.
- Exhaust gas emissions are in the limelight these days, and many countries have implemented regulations to limit emissions from ships. Non-compliance (NC) can lead to fines and penalties. Reducing fuel consumption and switching to cleaner fuels is the best way to comply with these regulations.
- Technological advancements will drive the effort to reduce fuel consumption as it dawns on ship-owners that they cannot keep operating

their ships as usual. For example, hybrid ships using renewable energy and alternate fuels will drive the move to reduce fuel consumption in the next two decades.

Thus, we see that reducing fuel consumption on ships has multiple benefits for all stakeholders.

It results in significant cost savings for operators, enhances environmental sustainability by reducing emissions, contributes to energy security by diversifying fuel sources from fossil fuels to other alternatives, and improves stakeholder relations by demonstrating a commitment to sustainability. We will be discussing this in detail in future chapters.

2.3.2 Installing Fuel-Efficient Engines

The exhaust gas from the engines contains a huge amount of heat energy. Fuel-efficient engines strive to utilise this heat energy to power their turbines, run steam plants, and produce electrical energy, thereby increasing the efficiency to almost 80%. As air pollution becomes a major factor, engine manufacturers work harder at new prototypes to reduce carbon footprint as well as the emission of noxious gases. For example, it is a known fact that nitrogen is an inert gas and combusts only at high pressure and high temperature for NO_x. Thus, intelligent engines tweak the firing cycle so that the fuel inlet is at off-peak pressures. Further water spray around the fuel ingress in the furnace reduces the temperature somewhat. These twin factors considerably reduce the combustion of nitrogen, bringing down the emission of NOX. As per MARPOL Annex VI, vessels have to maintain a technical file related to their nitrogen emissions. Regulation 13 of MARPOL Annex VI requires vessels delivered after 1 January 2016 to comply with the stringent Tier III limits for NO_x emissions which is possible by the new technologies available in engine design.

Fuel-efficient engines can improve energy efficiency on ships, benefiting both the maritime industry and the environment. By reducing fuel consumption, GHG emissions, and operating costs, fuel-efficient engines help make maritime transport more sustainable and economically viable in the long term.

2.3.3 Increasing the Overall Energy Efficiency of the Ship

On board, the ship's crew can take several measures to increase the energy efficiency of the ship and thereby reduce the fuel consumption:

- **Voyage Planning:** This is the most basic and important measure to reduce fuel consumption. The voyage plan should be the most economical route from one port to the other, commensurate with the safety of the ship. It is to be noted that adverse weather and currents will increase energy consumption. One of the most important aspects of voyage planning which will affect energy efficiency is the distance covered. A careful study of the voyage plan has to be conducted to decide where the distance steamed can be reduced. Keeping in mind the safety of the vessel, the voyage plan should be along the shortest distance from one port to the next.

By carefully planning the route, the most efficient path can be selected by considering factors such as ocean currents, weather conditions, and traffic congestion. This will lead to a reduction in fuel consumption and contribute to sustainable operations.

- **Weather Routing:** Most shipping companies employ the services of weather routing companies to guide the master on the most optimum course, keeping in mind the weather, currents, and the ship's requirements. Nowadays, weather routing services have resorted to artificial intelligence to obtain an accurate prognosis of weather patterns and thereby advise the vessel on the course and speed, keeping in mind the safety of the ship and the economics of the voyage. The master must be in communication with them and express any concerns or differences with their advice. However, it is to be noted that the master retains overriding authority in all aspects of ship operation. If the master feels that the route advised by the weather routing services is unsafe in any way, they should raise the red flag and inform the routing services accordingly.

  Avoiding adverse weather conditions and unfavourable currents can significantly reduce the fuel consumption of the ship. Weather routing improves energy efficiency in shipping by optimising vessel routes to minimise fuel consumption, reduce emissions, enhance safety, and improve operational efficiency.

- **Speed Optimisation:** Speed is an important criterion in controlling fuel consumption. A 10% increase in speed will increase fuel consumption by more than 15% depending on the size and type of ship. Thus, the speed of the ship is to be adjusted in case of berthing delays and so on. This is known as "just in time" arrival. For this to be successful, the ship's agents must coordinate with the port and the pilots to get an accurate estimate of the berthing of the vessel. They can then advise the master accordingly, who will then adjust their speed to arrive on time and thereby reduce fuel consumption.

  Speed optimisation balances vessel speed and fuel consumption, with techniques like slow steaming significantly reducing fuel usage without compromising timely arrival and by improving energy efficiency.

- **Trim of the Vessel:** Trim has been in the news for a long time as the jury is out on whether a trim by head or a trim by stern is better for the speed, fuel consumption, and energy efficiency of the ship. Ultimately, the right trim depends on various factors such as the design of the ship, the deadweight, the weather conditions, and so on. It is for the ship operators to study this aspect under different operating conditions to arrive at the right trim for the vessel, which will optimise the hydrodynamic performance and energy efficiency of the ship.

- **Ballast Water Management:** Ballast water management is in the news of late due to the ballast water management convention, which requires all vessels to phase in a BWT plant between 8 September 2019 and 7 September 2024. Proper management of ballast has an effect on the deadweight of the ship, her trim, as well as her energy efficiency. Although there is nothing much the ship staff can do in this regard, maintaining the minimum ballast on board will no doubt lead to less fuel consumption and subsequently reduce the carbon footprint of the vessel. This is not always possible as, in bad weather, ships may have to take on additional ballast to make the ship more stable.

- **Equipment Running Hours:** All equipment consumes energy, and cutting down on equipment running hours is a good method to increase energy efficiency. Whether it is the mooring equipment or the auxiliary equipment in the engine room, a little care to ensure that equipment does not run idle will help in reducing energy consumption. For example, it is a common practice to keep the mooring winch and windlass running well before anchoring or berthing operation. It is a good practice for vessels to develop standard operating procedures (SOPs) for equipment usage to streamline the running hours and ensure energy efficiency. By providing standardised guidelines and training, SOPs enable ship operators to optimise equipment performance, minimise fuel consumption, and reduce environmental impact while ensuring safe and efficient operations at sea.

- **Direct Energy Consumption on Board:** There are many other ways to reduce fuel consumption and increase energy efficiency, such as controlling the air conditioning to maintain the required temperature, controlling cargo heating, switching off lights when not required, and so on. The ship staff is directly involved in these operational measures, and managing and controlling the energy consumption on board will go a long way towards energy efficiency. The IMO has made the Ship Energy Efficiency Management Plan (SEEMP) a mandatory requirement on board ships.

- **Training and Motivation of the Ship Staff:** There is no doubt that any system is only as good as the people who implement these systems. With this in mind, training of ship staff is in focus. Although there is no mandatory training requirement for energy efficiency, many shipowners enrol their ship staff in such courses conducted by recognised organisations (ROs) and other well-known training centres. The staff are educated regarding the importance of energy efficiency and provided with various measures in order to implement the SEEMP and thereby reduce the GHG emissions from their ships.

- **ISO 14001:** The International Organization for Standardization (ISO) has various standards for various activities. The ISO is an international, non-governmental organisation that develops standards for the safety and efficiency of products and services. The ISO 14001 is the standard for environmental management. Shipping companies opt for the implementation of this standard both on their ships and in their offices. Before the relevant certification, audits are carried out by certification bodies approved by the administration. ISO 14001 is a useful standard for ship-owners to ensure that their ships follow the various requirements for environmental protection and energy efficiency.
- **Installing Scrubbers on Board Ships:** As per IMO's amendments to Annex VI, the sulphur limit for fuel used in the ship's engines is 0.5% m/m outside the ECAs and 0.1% m/m within the ECA. Ship-owners may, at their option, install scrubbers to remove the SOx from the exhaust gases and continue to use fuel with the 3.5% m/m as per the earlier regulations. Since the open-loop scrubbers discharge the wastewater containing the SOx into the sea, many countries have prohibited the use of open-loop scrubbers within their waters. Closed-loop scrubbers are a more sustainable option as the wastewater is discharged to reception facilities ashore.

These are strategies for increasing the fuel efficiency on board ships, and we shall discuss these in greater detail in later chapters. But what is important to remember is that shipboard energy efficiency is very important in the overall ambition to reduce energy consumption. To aid with this, the IMO has introduced the SEEMP, which assists ship-owners and ships in reducing fuel consumption by various measures as detailed in the plan. Further, IMO has introduced the concept of calculation of the EEDI and the energy efficiency operational index (EEOI). These mandatory measures will assist the ship-owners to take measures to reduce the energy consumption on their ships on a year-to-year basis. Let us have a quick look at these terms and what they mean:

- The EEDI is a standard approved by IMO for new ships. It is to ensure that these ships are energy efficient and have reduced GHG emissions. The value of EEDI is calculated for each individual ship as per IMO guidelines.
- Attained EEDI is the EEDI achieved by individual ships.
- Required EEDI is the maximum value of EEDI for a ship based on the ship type and size.
- The reference line is the average EEDI for the same ship type determined by the IMO using data from vessels constructed between 1999 and 2008.
- The EEXI is used to assess the energy efficiency of ships already in service.

- The EEOI was developed by the IMO in order to allow ship-owners to measure the fuel efficiency of a ship in operation.
- The carbon intensity indicator (CII) is a measure of the carbon footprint of the ships and is given by the grams of CO_2 emitted per transport mile.

All these terms will be explained in detail in the next chapter.

2.4 SUMMARY

Every effort should be made on board to increase energy efficiency and reduce fuel consumption. Only then shall the shipping industry reduce its carbon footprint and contribute to a cleaner and healthier world. In this chapter, we discussed the current state of sustainability and the harmful effects of shipboard emissions. We followed this up by briefly discussing some methods to address sustainability issues, such as compliance with MARPOL Annex VI, as well as improving the energy efficiency of the ship to reduce fuel consumption and GHG emissions.

We will be discussing the various aspects of energy efficiency in future chapters. The next chapter will delve into the IMO measures to reduce GHG emissions from ships by implementing the new requirements of SEEMP and other aspects of MARPOL. It is important for stakeholders in the shipping industry to take note of these measures and implement them on board ships so that GHG emissions are reduced by reduction of fuel consumption. A knowledge of energy efficiency will go a long way in meeting the stringent requirements and ambitions of IMO in this regard.

BIBLIOGRAPHY

4. Statista. (2023). *Carbon Dioxide Emissions of the Most Polluting Countries Worldwide in 2010 and 2022.* https://www.statista.com/statistics/270499/co2-emissions-in-selected-countries.

   Statista is a global data and business intelligence with a collection of statistics. Intensive study has been conducted on the carbon dioxide emissions country-wise which will help us to determine the top emitters worldwide.

5. International Maritime Organization (IMO). (2021). *Fourth IMO GHG Study 2020.* https://wwwcdn.imo.org/localresources/en/OurWork/Environment/Documents/Fourth%20IMO%20GHG%20Study%202020%20Executive-Summary.pdf.

   The fourth GHG study conducted by IMO reveals many interesting statistics, including the CO_2 emissions from shipping, the emission projections, and other data related to emissions from ships.

Chapter 3

Energy Efficiency Indices and SEEMP

The Ship Energy Efficiency Management Plan (SEEMP) is a requirement of Annex VI of MARPOL for vessels to reduce greenhouse gas (GHG) emissions and increase energy efficiency.

The SEEMP outlines a range of actions and measures to assist the ship-owner in enhancing fuel efficiency and reducing the carbon footprint of the ship. SEEMP helps ship-owners meet their obligations towards environmental sustainability

The International Maritime Organization (IMO) has been continuously working to limit the exhaust gas emissions from ships. Although emissions from ships constitute a small percentage of overall emissions, they are bound to increase exponentially if unchecked. Thus, the focus of IMO is to ensure that each ship takes adequate measures to reduce its carbon emissions. SEEMP is the main requirement regarding this.

In this chapter, we will discuss the SEEMP, which is a requirement of MARPOL Annex VI. We also explain the implementation of SEEMP on board. Annex VI includes provisions for both the Energy Efficiency Design Index (EEDI) for new ships and the Energy Efficiency Existing Ship Index (EEXI) for existing ships.

Before proceeding further, let us discuss the various indices for measuring the GHG emissions from ships:

- The EEDI is a measure to assess the efficiency of a ship's design. IMO's MEPC resolution MEPC 364(79) details the guidelines for the calculation of the attained EEDI for new ships. The EEDI sets specific efficiency targets based on the ship's size, type, and capacity, aiming to encourage the development and construction of more energy-efficient vessels. EEDI is a standard for sustainable and energy-efficient practices on ships.
- The EEXI is another standard to improve the energy efficiency of existing ships. Based on the technical and operational parameters of the ship, the EEXI is calculated. The implementation of EEXI and the compliance by existing ships is part of the IMO's strategy to reduce GHG emissions from ships.

DOI: 10.1201/9781032702568-3

- The Energy Efficiency Operational Indicator (EEOI) is a measure of the actual performance of the ship. EEDI and EEXI are reference values, but the EEOI is based on the actual fuel consumption of the ship related to the transport work.
- CII is a rating depending on the fuel consumption of the ship. The attained CII will be calculated as grams of CO_2 per tonne-mile based on the fuel consumption for the year from 1 January to 31 December 2023. Accordingly, a CII rating will be issued to the ship. In order to enable a correct estimate of the CII, the data collection system (DCS) was established by IMO as a mandatory requirement from 1 January 2019 for administration or RO. To facilitate this, ships have to submit a data collection plan (DCP) specifying the methods of collection and reporting of their fuel consumption.

These will be discussed in detail later on in this chapter. Now, let us look at what exactly SEEMP is.

3.1 UNDERSTANDING THE FRAMEWORK AND STRUCTURE OF THE SEEMP

The SEEMP is a set of guidelines for implementation on board with the aim of improving the energy efficiency of the ship as well as optimising energy usage. Its basic purpose is to assist the ship's crew and the ship operators to reduce fuel consumption and control GHG emissions through energy efficiency.

The responsibility for preparing the SEEMP lies on the ship-owner, operator, or the ship management company, that is, the entity responsible for the operations of the ship. They may consult with technical experts or consultants for guidance. Classification societies and RO often offer guidelines and may assist in the development of the SEEMP.

The SEEMP framework consists of the following elements:

1 **Introduction**: This contains the purpose and objective of the SEEMP and establishes the groundwork for the implementation of the measures to reduce energy consumption. It will also contain the regulatory requirements, including IMO, national, and international requirements as applicable.
2 **Company Policy**: This contains the vision of the company towards energy efficiency and contains a statement highlighting the ship-owner's commitment and environmental responsibility.
3 **Operational Measures**: This is the most important aspect of the SEEMP and contains the operational measures to increase energy efficiency. These measures include voyage planning, speed optimisation, weather routing, as well as optimising the trim and draft of the vessel.

Effective use of shipboard auxiliaries and equipment and engine load management are also the measures generally included in the SEEMP for energy efficiency.

4 **Technical Measures:** Machinery and equipment which are not maintained will not be efficient. Thus proper inspection and maintenance procedures are necessary to ensure the optimal performance of machinery and equipment. Further this section contains measures to improve energy efficiency such as regular hull cleaning, adopting advanced technologies, and so on.

5 **Monitoring and Reporting:** As per MARPOL Annex VI, ships of 5,000 GT and above have to report their fuel oil consumption data to their flag administration after the end of each calendar year. The flag then transfers this data to the IMO database.

The SEEMP details ship-specific measures to increase the energy efficiency of the ship. These measures include the use of alternative fuels, monitoring and reporting, crew training, and data analysis to identify trends and opportunities for further improvement. This responsibility is equally shared by the company which has overall responsibility for the ship operations. Thus, from the development and implementation to the maintenance of the requirements of SEEMP, the company has to support the vessel and provide all necessary support so that the ship can improve her energy efficiency and reduce her carbon footprint.

Now let us have a look at the various components that make the SEEMP as well as the various information contained therein.

3.2 BREAKING DOWN THE SEEMP

As per Annex VI of MARPOL, the requirement for SEEMP entered into force on 1 January 2013. Here are some of the key features of SEEMP:

- It is a ship-specific document which contains energy efficiency measures identified by the ship-owner or manager.
- It is a tool for monitoring ship and fleet energy efficiency.
- It requires the ship-owner to adopt best management practices and new technologies to make their ship energy efficient.

SEMP shall contain the following information:

- Best practices for saving energy.
- Voluntary operational index. This shall be calculated for each voyage and for voyages over a specific time period. The specific calculations for these indices will be covered later on in this chapter.
- Other voluntary operational measures which the companies are required to formulate and document.

- Energy inventory which specifies the utilisation of energy on board the ship, so as to obtain a reasonable estimate of energy consumption.

SEEMP was first implemented on ships as a voluntary measure in January 2013. But prior to that, the details of implementation were not clear as it was not yet mandatory. Around that time, I was imparting value-added training to the sailing officers of a ship management company. As part of the training, I was supposed to give the officers a briefing and familiarise them with SEEMP Part, but the details were neither clearly specified nor easily available. Through extensive studies and communication with classification societies, I was able to structure a familiarisation module on SEEMP as part of the QMS training.

3.2.1 SEEMP Part I

Along with the SEEMP, the EEDI also entered into force on 1 January 2013. SEEMP was the initial document dealing with on-board energy efficiency and states the company's line of action to follow the requirements of MARPOL Annex VI. No approvals were required initially, but the ship had to maintain documentation as evidence of their compliance with the regulations. Later on, it came to be known as SEEMP Part 1 when the need for an upgradation to Part 2 was required.

SEEMP Part 1 should contain the guidelines to improve the ship's energy efficiency through a structured approach involving four crucial stages:

- **Planning:** This stage involves outlining the measures to improve the ship's energy efficiency related to voyage planning, speed optimisation, and other relevant criteria to reduce fuel consumption.
- **Implementation:** This is the process of putting the planning into action. The energy efficiency measures, as detailed in the planning stage, will have to be incorporated into the vessel's operations.
- **Monitoring:** This stage involves tracking the vessel's energy efficiency and assessing the ship's performance after the implementation of the energy-saving measures.
- **Self-Evaluation and Improvement:** This step is crucial as it involves self-assessment to determine the effectiveness of the measures implemented. It includes audits, feedback, learning from experience, and adjusting the strategy as required.

SEEMP Part 1 should contain the following:

1 Ship's particulars
2 Measures to improve energy efficiency
3 Monitoring tools for same
4 Measurable goals for reduction of energy consumption
5 Evaluation procedures such as audits, feedback, and so on

3.2.2 SEEMP Part II

MARPOL's new regulation 22A came into effect on 1 January 2019. Known as IMO's DCS, ships of 5,000 GT and above are required to collect their fuel consumption data and report to their flag state or their recognised organisation. A Certificate of Compliance (CoC) will then be issued to the vessel.

SEEMP Part II requires approval by the flag state or the RO and should be available on board prior to the date of entry into force (1 January 2019).

From 1 January 2019, ships should start collecting the required data as per Appendix IX of MEPC 278(70) and submit it to the administration for each calendar year or part of the year.

As part of the DCP, the following data is to be collected by the ship staff and reported to the ship-owner or operator:

1 IMO number
2 Ship type, gross tonnage, net tonnage, DWT
3 Power output, EEDI values
4 Start and end date of report
5 Fuel types, consumption during the period, and methods of collecting the same.
6 Distance travelled and hours underway.

The above information for the year will have to be sent by the ship-owner or operator to the administration or RO by the 31 March each year. Once the data is received and verified to be in compliance with regulation 22A, a Statement of Compliance will be issued valid for the following year.

Within one month of receiving the data, the administration or RO will transfer the same to the IMO ship fuel oil consumption database.

SEEMP Part II should contain the following:

1 Ships particulars
2 Fuel oil types and their consumption
3 Emission factor of fuel used (as mentioned in the 2014 guidelines)
4 Methods to measure fuel consumption
5 Distance travelled, hours underway, and methods for measurement of the same
6 Process used for reporting the DCP to the administration or RO
7 Statement of Compliance for the year issued by the administration after verifying the fuel oil data collection method and the actual data submitted
8 Copy of the DCP

3.2.3 SEEMP Part III

In order to ensure that the vessel complies with current MARPOL regulations, IMO has introduced the SEEMP Part III. All vessels must be in possession of

SEEMP duly approved by the flag administration by 1 January 2023. Verification of this can be done by the RO on behalf of the flag administration.

The timeline for calculating and reporting the CII as envisaged in SEEMP Part III (we will discuss CII in detail later in this chapter):

1 1 November 2022: MARPOL Annex VI related to EEXI and CII measures enters into force prior to 1 January 2023.
2 Prior to 1 January 2023: SEEMP Part III compliance with MARPOL Annex VI, Reg 26.3.1 to be verified (by RO)
3 Prior to 1 April 2024: Attained operational CII to be calculated and reported
4 Prior to 1 June 2024: Reported attained. 2023 operational CII to be verified 2023 operational carbon intensity rating in compliance with Reg 25 to be determined.

SEEMP Part III should include the following:

1 Methodology used to calculate the vessel's annual operational CII
2 Required annual operational CII for the next three years
3 The targeted CII for the next three years based on the individual ship's performance
4 Details on how this reduction in operational CII is to be achieved for the next three years
5 Self-evaluation and improvement
6 Correction plan in case the vessel receives an inferior rating

The guidelines for developing SEEMP are detailed in *IMO's 2022 Guidelines for the Development of a Ship Energy Efficiency Management Plan (SEEMP)*.[6]

Now, let us discuss the various means and methods to implement the SEEMP. Any plan is as good as its implementation, and this is the reason ship's staff must pay careful attention to the requirements of the SEEMP and ensure it is implemented.

3.3 IMPLEMENTING SEEMP

Implementation of an effective SEEMP involves procedures to put into practice the energy efficiency measures outlined in the plan. Specifically, it includes the following:

1 **SEEMP Development and Understanding:** Before implementing a SEEMP, it should be ensured that the plan is well developed, and all the relevant stakeholders understand its content and objectives. This may involve training and education for the crew and staff. Even the shore

staff, such as superintendents, must be made aware of the intricacies of the SEEMP so that they can fully implement all the requirements.

2 **Leadership and Commitment:** Ensure that senior management and ship-owners are committed to the SEEMP's implementation. Their support is critical for allocating resources and fostering a culture of energy efficiency. It is often members of the senior management who resist change and prefer to stick to the old procedures. The top management must make it amply clear that there is no scope for laxity as far as the implementation of SEEMP is concerned.

3 **Assign Responsibilities:** Clearly define roles and responsibilities for SEEMP implementation. Designate individuals or teams responsible for overseeing and executing the plan. Establish an energy efficiency officer or coordinator role if necessary. It is by specifying levels of authority and responsibility that individuals assigned with specific tasks will make the extra effort in compliance. If this is not done, then laxity will set in and the entire exercise becomes futile.

4 **Crew Training and Awareness:** Any system is as good as the personnel who work towards its implementation and continuous improvement. Thus, it is necessary to train the ship's crew and staff on the SEEMP and the specific energy efficiency measures it contains. The management must ensure that everyone understands their roles and responsibilities in implementing these measures.

5 **Data Collection and Monitoring:** As part of MARPOL's requirements, a system has to be set up for collecting relevant data, such as fuel consumption, emissions, and energy-efficient practices. This may involve installing monitoring equipment or using software systems to track and analyse data. These data have to be collated and forwarded to the flag administration who will then forward the consolidated data to the IMO.

6 **Key Performance Indicators:** Specific key performance indicators (KPIs) should be defined to measure the effectiveness of the SEEMP. These KPIs should align with the goals and objectives of the plan and provide a clear way to assess performance. Once the KPIs are defined, the ships will have to follow them and document their compliance with the required performance parameters.

7 **Regular Audits and Inspections:** A system of regular audits and inspections should be set up to ensure compliance with the SEEMP. Auditors can assess whether the energy efficiency measures are being followed and whether they are achieving the desired results. They may raise NC in case they are not satisfied that the vessel is complying with SEEMP. In such a scenario, corrective actions have to be taken and the NC closed as soon as possible.

8 **Documentation and Reporting:** Compliance with any system can be maintained by documenting as well as through proper inspections, audits, and reporting to the top management. Thus, detailed records of

SEEMP implementation have to be maintained, including data, audit reports, and corrective actions taken. Further, regular reports on the ship's energy efficiency performance have to be made to the relevant authorities and stakeholders, as required.

9 **Regulatory Compliance:** The requirement of implementing and maintaining an effective SEEMP is as per MARPOL Annex VI. It has to be ensured that the SEEMP implementation system aligns with international and national regulations related to energy efficiency in the maritime industry. The ship operators will have to consult these requirements and regulations at the time of developing the SEEMP.

10 **Periodic Review:** In order to ensure continuous improvement, it is necessary to periodically review the SEEMP and the implementation system. The review must be carried out by both the concerned ship staff and the relevant shore staff in order to confirm that the SEEMP is being followed, and it is updated with no changes in technology, regulations, and industry best practices.

Next, let's look at why these measures should be implemented.

3.4 THE IMPORTANCE OF IMPLEMENTING THE SEEMP

We have seen that the SEEMP is a mandatory requirement as per MARPOL Annex VI. It then comes within the ambit of the port state control and flag administration. A ship has to demonstrate compliance with these requirements or be subjected to penalties and ship detentions. Looking beyond penalties, ship-owners would prefer to abide by these requirements for commercial reasons or even for ethical reasons. Whatever the reason may be, it is an established fact that ships need to improve their energy efficiency, leading to a reduction in fuel consumption and thereby GHG emissions.

Once a ship and a company have identified the measures to implement the SEEMP, establishing a systematic approach to implementing these selected measures is crucial. This involves developing energy management procedures, defining tasks, and assigning them to qualified personnel. Consequently, the SEEMP should outline how each measure will be implemented and designate responsible individuals. The effectiveness of these measures should be carefully checked in order to ensure continuous improvement.

Let's look at a real-life example to illustrate the importance of SEEMP. In March 2023, an 18-year-old bulk carrier was boarded by the port state control in Zeebrugge, Belgium for a routine PSC inspection. As part of their documentation check, they found substantial shortfalls in the vessel's SEEMP, as follows:

*   The crew were not aware of the SEEMP and the actions required to maintain an effective SEEMP. Only the senior officers had some

idea of energy efficiency and the operational measures required for implementation.

- There was no record of any audit being carried out to determine short-falls in the system. If an audit had been carried out, corrective actions could have been taken to make the SEEMP more effective.
- The master and chief engineer could not demonstrate leadership quali-ties to ensure that the SEEMP was being followed. This led the crew to believe that it was yet another document without any real relevance on board.

The port state control took serious note of these deficiencies and detained the ship in port till corrective actions were taken. What we can take away from this incident is that vessels need to implement SEEMP seriously and maintain proper records of crew training, audits, documentation, and so on to ensure that they are in compliance with mandatory requirements.

Let us now have a look at the methods of measuring GHG emissions, namely the EEDI and the EEXI. These two indices will help measure whether the ship is complying with the emission regulations.

3.5 AN OVERVIEW OF EEDI AND EEXI

IMO has introduced two new indices to control and monitor energy effi-ciency on board ships. These are the EEDI and the EEXI. The EEDI has been in force for ships delivered on or after 1 July 2015, while the EEXI is mandatory for existing ships from 1 January 2023. Let us have a closer look at these two indices.

3.5.1 EEDI for New Ships

It has long been established that for a realistic reduction of GHGs from a ship's exhaust, there should be some modifications at the design stage of the ship. With the help of classification societies and other technical organisa-tions, the IMO worked on this aspect and brought about the EEDI. This is a design mechanism for new ships to help them attain the required energy efficiency level. The choice of technology to be used to achieve this is left to the industry in general and the engine manufacturers in particular to adopt. The ship designers and builders will be free to use the best solutions to com-ply with the regulations. The EEDI standards will be phased in from 2013 to 2025 as new ships are being delivered. Vessels are required to carry a class-approved certificate on EEDI. Shipyards are responsible for calculating the EEDI of the ships, which is then verified by the classification societies. A tech-nical file has to be prepared by the shipyard containing the EEDI calculation.

Attained EEDI is the value achieved on any ship. It is ship-specific and calculated based on the data in the ship's EEDI technical file.

Required EEDI is the maximum permitted value of attained EEDI, dependent on the type of ship and its size.

MARPOL Annex VI makes it mandatory for all vessels of 400 gross tonnage and above to comply with the EEDI regulations as follows:

- A ship whose building contract is placed on or after 1 January 2013.
- A ship whose keel is laid on or after 1 July 2013.
- A ship whose date of delivery is on or after 1 July 2015.

Attained EEDI can be defined as the "environmental cost" divided by the "benefit for society". This can be expressed as

EEDI = Impact to environment / Benefit for society

The impact to environment is the CO_2 emission, while the benefit to society is the cargo transported from one place to the other.

Attained EEDI can thus be given by

EEDI = CO_2 emission / Transport work

where
- the CO_2 emission is the total CO_2 emission from combustion of fuel, whether from main propulsion or the auxiliary engines and
- the transport work is the ship's capacity (dwt) * ship's design speed measured at the summer load condition and at 75% of the rated installed shaft power.

Although the above formula looks relatively simple, calculating the EEDI is a complex process, as it involves several variables and considerations. More specifically, if we consider these variables, the formula for EEDI can also be calculated as

EEDI (in grams of CO_2 per tonne-mile) = $(C \times DWT) / (V \times L \times \beta)$

where
- "C" is the estimated carbon dioxide (CO_2) emissions in total from the main and auxiliary engines and boilers of the ship over a specific reference period, expressed in metric tons.
- "DWT" is the deadweight tonnage of the ship, which represents the cargo-carrying capacity, expressed in metric tons.
- "V" is the design speed of the ship in knots.
- "L" is the length between perpendiculars, representing the overall length of the ship, expressed in metres.
- "β" is a factor that accounts for the ship type and size and is specific to each ship category.

The required EEDI for ships is calculated using the reference line (baseline), which represents an average index value. Reference lines are set up for various ship types. The IMO resolution MEPC 231(65) 2013[7] gives the methodology for calculating the reference lines:

Required EEDI = (1 − X/100) × EEDI reference line

The IMO guidelines for EEDI calculations are explained in *IMO Resolution MEPC.364(79)* for the 2022 guidelines for EEDI calculations.[8] *IMO Resolution MEPC.324(75)* guidelines for required EEDI[9] specify these details.

The benefits of EEDI will unfold gradually as older ships are scrapped and new ships with the required EEDI are delivered. As per the IMO regulations a country can delay the implementation of EEDI by four years without any penalty. Considering the average age of ship to be 25 years, it can be assumed that by 2040 to 2045, most of the ships would be EEDI compliant.

The International Council on Clean Transportation has estimated that the EEDI regulations would save approximately 15–45 million metric tons (mmt) of CO_2 annually by 2020 and between 141 and 263 mmt of CO_2 annually by 2030.

3.5.2 How Can New Ships Ensure Compliance With EEDI Requirements?

There is no doubt that design modifications are the most important aspect of EEDI. Shipbuilders will have to find various ways and means to reduce GHG emissions so that their EEDI remains below the baseline. The baseline is the reference for the acceptable EEDI. If a vessel's EEDI is above the baseline they will be excluded from the market and will have serious trouble finding cargo and charterers. Shipbuilders will have to calculate the attained EEDI and ensure that it is below the baseline of the required EEDI, if required by reducing the design speed or by increasing the energy efficiency of the ship by various means.

The following technologies are being used to comply with the EEDI requirements:

- **Hull Design:** A streamlined hull with reduced wave-making resistance is the most effective way to meet the EEDI requirements. Shipyards constantly work on fine-tuning the hull form or incorporating hydrodynamic features.
- **Hull Coatings:** The use of high-performance anti-fouling paints, such as silicon paints, can considerably reduce marine growth and associated frictional losses.
- **Propellers:** Shipbuilders have long recognised that propellers are an effective tool to reduce GHG emissions by reducing turbulence and drag. Thus, they constantly experiment with new types of propellers.

Contra-rotating propellers, Kort Nozzle propellers, Azimuth propellers, controllable pitch propellers, twin screws, and so on are all effective to a certain extent depending on the size of the ship and the requirements of the ship-owners. Needless to say, cost is an important factor in deciding on the type of propeller for individual ships.

- **Engine Designs:** Marine engine manufacturers invest in new technologies to come up with fuel-efficient solutions. Intelligent engines with automation have been found to increase fuel efficiency. Some manufacturers use heat recovery systems and secondary cycles to further increase efficiency and reduce fuel consumption. The scope for fuel-efficient engines to meet the required EEDI is endless, and as new technologies are introduced, the fuel efficiency of marine engines will improve.
- **Auxiliary Equipment:** Ventilation and air conditioning on ships, mooring equipment, and other auxiliary devices consume energy and by making them energy efficient, fuel consumption can be reduced.
- **Hybrid Ships:** This concept is picking up, although hybrid vessels are being produced on a smaller scale. These ships are capable of propulsion by renewable sources as well as by conventional fuels. They have automation that can allow switching from one source to the other, or a combination of both to achieve maximum efficiency. For example, a hybrid vessel with wind energy will switch over to the conventional fuel if the wind is unfavourable or insufficient.
- **Data Analytics:** Artificial intelligence and machine learning are increasingly being used in ship design to analyse data, identify patterns, and come up with energy-efficient solutions to reduce EEDI.

We will be discussing these in greater detail later on.

As per IMO's EEDI guidelines, the following are to be considered for new build ships:

- New building ships must be more energy-efficient than the baseline, thus reducing their carbon intensity.
- Performance targets are increasingly stringent over time, thus incentivising innovation in ship design is required.
- There are different goals for different ships, hence the specifications for different types of ships must be taken into account. For example, large container vessels (DWT greater than 200,000 tonnes) built after 1 April 2022, must be 50% more efficient than the baseline.

The requirements for EEDI get more stringent with time. There are several phases of EEDI, shipyards will have to follow the phase that fits their timeline of vessel delivery. The following are the different phases:

- Phase 0: In force currently for ships delivered on or after 1 July 2015 and before 1 January 2019.

- Phase 1: For ships delivered on or after 1 January 2019 and before 1 January 2024.
- Phase 2: For ships delivered on or after 1 January 2024 and before 1 April 2026
- Phase 3: For ships delivered on or after 1 April 2026.

EEDI is applicable to all ships greater than 400 GRT trading in international waters, except as follows:

- Those operating only in their flag's national waters
- Ships not propelled by mechanical means
- Floating production storage and offloading (FPSO), floating storage units (FSU), and drilling rigs

The EEDI took care of CO_2 emissions from new ships, which were fitted with new technologies to reduce GHG emissions. But what about existing ships? What steps did IMO take to control GHG emissions for existing ships? This is where the EEXI comes in.

3.5.3 EEXI for Existing Ships

The MARPOL Annex VI amendments entered into force on 1 November 2022. As per these amendments all ships have to calculate their attained EEXI effective 1 January 2023. In addition, they have to initiate the DCS for the reporting of their annual operational CII and CII rating.

The attained energy efficiency index is calculated and then compared against the required EEXI. If the attained EEXI is more than the required EEXI it would indicate that the vessel has to take further measures to reduce her EEXI.

The required EEXI is calculated using the EEDI for the ship as a reference. A reduction factor is applied depending on the ship type and size. This is given by

Required EEXI = (1 – X/100) × EEDI reference line

Where X is the reduction factor which depends on the ship type and size.

The required EEXI is calculated and compared against the attained EEXI to measure the vessel's compliance with MARPOL Annex IV and relevant IMO resolutions.

The attained EEXI is calculated using the EEDI as the baseline. We have already seen the method of calculation of EEDI. The EEXI is calculated as follows:

EEXI = (EIV – (P1 * D1 + P2 * D2 + P3 * D3)) / TEF,

where
- EIV is the existing EEDI as calculated.

- P1, P2, and P3 are parameters based on the ship's age, size, and the type. These parameters ensure that the EEXI requirements are specific to the vessel.
- D1, D2, and D3 are reduction factors to be achieved through technical and operational measures. These values depend on the actual improvements made to the vessel. These deduction values are important for calculation of EEXI as ships which have taken measures towards energy efficiency improvements will be rewarded with a lower EEXI.
- TEF stands for "time of exposure factor". This factor accounts for the remaining operating period of the ship in years. Thus, if the TEF is shorter, the EEXI is adjusted upwards requiring more stringent energy-efficient improvements. A shorter TEF would indicate that the vessel is at the end of its operational life, and energy efficiency may not be of the highest standards. Thus a higher EEXI is allocated on the basis of the calculations.

The equations provided earlier have been simplified for ease of understanding. The details can be understood in the report on the EEDI issued by the Classification Society Indian Register of Shipping (IRClass).[10]

The calculations for EEXI are quite complex, and the details are given in the documentation on the IMO Resolution MEPC.350(78).[11]

3.5.4 Comparing the Required EEXI and the Attained EEXI

The attained EEXI should be lower than the required EEXI. A lower EEXI for ships is beneficial for several reasons, both for the shipping industry and for the environment:

- **Regulatory Compliance:** A lower EEXI indicates that a ship complies with the IMO's energy efficiency regulations. Non-compliance may mean penalties and fines.
- **Reduced Fuel Consumption:** Ships with lower EEXI values are generally more energy-efficient and consume less fuel for a given amount of cargo transported, resulting in cost-benefit and reduction of the carbon footprint of the vessel.
- **Industry Acceptance:** Since vessels with lower EEXI are eco-friendly and emit lesser GHG, they are more acceptable to the industry, resulting in a competitive advantage. The company's reputation would be enhanced and charterers, shippers and customers in general would find these vessels more attractive.
- **Port Entry:** Some ports would give preference to vessels with lower EEXI reducing port delays and associated costs.

These advantages would spur the ship-owner to take all necessary measures to lower the EEXI of their vessels and contribute to lowering GHG emissions worldwide.

We are now familiar with the EEDI of new ships and the EEXI of existing ships. But what about the actual performance of the ship during its operational life? This is where the EEOI comes in, which measures the actual fuel consumption of the ship and thereby its energy efficiency.

EEOI is a function of the actual fuel consumption, the cargo carried, and the speed of the ship.

3.6 INTRODUCING THE ENERGY EFFICIENCY OPERATIONAL INDICATOR

The IMO has introduced the EEOI as a means for ship-owners to gauge a vessel's fuel efficiency during its active operation. EEOI serves as a CII and reflects the energy demand required for transporting goods. EEOI, calculated by dividing *annual fuel consumption* by *transport work*, provides an annual average representation of a ship's carbon emissions under its real operating conditions, accounting for various factors such as actual speeds, drafts, cargo capacity, distance travelled, and the impact of hull and machinery wear and tear, as well as weather conditions. Ship managers can employ EEOI as a tool to evaluate the impact of operational changes on their vessel's fuel efficiency, potentially prompting actions such as more frequent propeller cleaning, the adoption of new propellers, or the implementation of waste heat recovery (WHR) systems. It's worth noting that EEOI differs from the EEDI in that EEOI focuses on actual fuel consumption during a voyage, while EEDI is based on vessel design and theoretical emissions.

The objectives of EEOI are as follows:

- Assessment of energy efficiency during each voyage
- Appraisal of operational effectiveness by owners or operators
- Ongoing tracking of individual vessel performance
- Evaluation of alterations to the ship or its operations
- Measurement of energy efficiency during each voyage
- Appraisal of operational effectiveness by owners or operators
- Continuous monitoring of individual vessel performance
- Evaluation of alterations to the ship or its operations

EEOI can be given by

EEOI = (Fuel consumed × conversion factor) / (Cargo carried × distance travelled)

Since the emission of CO_2 from ships is directly related to the bunker fuel oil consumption, the EEOI can be considered to be a measure of the ship's fuel efficiency. The EEOI is a continuous observation of operational ships and can assess outcomes resulting from modifications to the ship or its operations. For instance, the impact of hull cleaning or retrofitting a more efficient propeller can be gauged through changes in the EEOI value. The implementation of operational measures such as just-in-time planning or advanced weather routing systems will result in improving the fuel efficiency, which can be observed in the EEOI value.

The general IMO guidelines on improving the energy efficiency of ships[12] specify the general methods to improve the technical performance of newly built ships and thereby meet their EEDI obligations. These include the following:

- Making them more energy efficient than the baseline, thus reducing their carbon intensity
- Incentivising innovation in ship design and lastly by having different goals for different ship types. For example, large container ships built after 1 April 2022 must be 50% more efficient than the baseline and so on.

3.7 BRIEF RECAP

Confused? It is quite normal to be confused with so many indices. Let us recap for the sake of simplicity:

The EEDI is a mandatory requirement on new ships to make them energy efficient, by means of incorporating latest technology and design.

$$EEDI = CO_2 \text{ emission/transport work}$$

The EEXI has been introduced to assess and improve the energy efficiency of existing ships. It uses the EEDI as a reference and applies reduction factors such as the ship's age, size type, and any new technological or operational measures by the ship.

The EEOI is a measure of the actual fuel consumption of the vessel during its operations and is used to gauge the real-time energy efficiency of the ship.

The EEDI, EEXI, and EEOI are indices to measure the energy efficiency of the ship. But directly correlated to this is the CII, which is a measure of fuel consumption and thereby the energy efficiency.

3.8 CARBON INTENSITY INDICATOR

It is well known that carbon dioxide is directly responsible for global warming. It is released into the atmosphere when fossil fuels are used to operate the ship's engines. CO_2 emissions from ships' exhaust are due to the fuel consumption as well as the quality of combustion. If the ship's engines are well maintained, then the combustion would be proportionately good, and the emission of CO_2 would reduce. The second factor affecting CO_2

emission is the speed of the ship. There can be no argument for the fact that ships proceeding at 24 knots will consume much more fuel than the same ship at 16 knots.

Thus, measuring the carbon intensity of the ship can give a true picture of the harmful effects of the ship's emissions. To achieve this, IMO has introduced the CII.

CII is an operational indicator of the ship's carbon footprint and is given in grams of CO_2 emitted per cargo carrying capacity and distance travelled (transport work). It is applicable to ships of 5,000 GT and above.

From 1 January 2023, ships have to report their fuel consumption data to the flag administration or their RO in order to calculate and report their annual operational CII and CII rating.

The MEPC resolution 336(76), which was adopted on 17 June 2021, details the guidelines on operational CII and their calculation methods:

- The required CII is the value that the ship must adhere to. It is calculated taking into account the reference CII and the reduction factor. The required CII is calculated by the formula:

  CII = (Annual fuel consumption × CO_2 emission factor) / (Distance sailed × DWT)

  There are several correction factors, depending on the type of ship and other ship parameters, as described in the official documentation for the IMO resolution MEPC.355(78)[14].

  Based on this formula, ships have to calculate their attained CII at the end of 2023.

- The attained CII of the ships can be taken as M/W, where M is the mass of CO_2, which is the sum of CO_2 emissions (in tons) for all the fuel oil consumed on board the ship in a given calendar year. It is the sum of the total fuel consumed in the year and the fuel mass to CO_2 conversion factor. W is the transport work, which is the product of the ship's capacity and the total distance travelled.

  Every ship of 5,000 GT and above shall, after the end of calendar year 2023 and the end of each following calendar year, be required to calculate their attained CII over the 12-month period of the calendar year. This is to be reported to the Administration within three months after the end of the calendar year.

Based on their performance vis-à-vis the required CII, the ships will be awarded an environmental rating as follows:

- A: Major superior
- B: Minor superior
- C: Moderate
- D: Minor inferior
- E: Inferior

Those ships with a D rating for three years or with an E rating will have to submit a corrective action plan and improve their CII by the next year. In addition, commercial pressures are building on ship-owners to ensure that they minimise the carbon footprint of their ships. In fact, oil majors and dry bulk charterers are beginning to go through the energy efficiency of the ships and their CII rating before fixing the charters. This will be a big incentive for the ship-owners to ensure that their ships are compliant with the regulations and are energy efficient. These ratings will become more stringent as 2030 approaches.

Guidelines on operational CII calculation are detailed in IMO MEPC Resolution 336(76), adopted on 17 June 2021.[14]

3.9 SUMMARY

Today, ship-owners have nowhere to hide. Due to the implementation of the measures described in this chapter, the days when old ships, with inefficient engines, belching smoke and causing both air and ocean pollution are over. Today it is easy for the port state control and the flag administrations to identify and blacklist defaulters. Not only would they be detained in port and subjected to hefty fines, but they would also face difficulty in finding cargo for their ships. Further, with the advent of EEDI, EEXI, EEOI, and CII, it becomes easier to measure whether the vessel is indeed complying with the requirements. It is important for ship-owners and operators to be aware of these indices and comply with the requirements.

In the next chapter we will discuss the general strategies to increase the ships energy efficiency. This will include IMO's revised GHG strategies and ambitions as well as the energy management systems adopted on board. The best practices for energy efficiency will also be highlighted.

BIBLIOGRAPHY

6. Resolution MEPC.346(78). (Adopted on 10 June 2022). *2022 Guidelines for the Development of a Ship Energy Efficiency Management Plan (SEEMP)*. https://wwwcdn.imo.org/localresources/en/KnowledgeCentre/Indexof IMOResolutions/MEPCDocuments/MEPC.346(78).pdf.

   This resolution details the guidelines for developing the SEEMP on board the ship. It includes the best practices for fuel-efficient ship operations, as well as the methodology for collecting fuel oil consumption data. The resolution also deals with the requirement of the CII as well as the plan for corrective actions for those vessels with inferior CII.

7. Resolution MEPC.231(65). (Adopted on 17 May 2013). *2013 Guidelines for Calculation of Reference Lines for Use With the Energy Efficiency Design Index (EEDI)*. https://wwwcdn.imo.org/localresources/en/KnowledgeCentre/ IndexofIMOResolutions/MEPCDocuments/MEPC.231(65).pdf.

The reference line can be calculated as given in this resolution, based on which the attained EEDI can be calculated.

8. Resolution MEPC.364(79). *2022 Guidelines on the Method of Calculation of the Attained Energy Efficiency Design Index (EEDI) for New Ships.* https://wwwcdn.imo.org/localresources/en/KnowledgeCentre/IndexofIMOResolutions/MEPCDocuments/MEPC.364(79).pdf.

The guidelines for the method used to calculate the EEDI give us the formula for calculating the EEDI for new ships. This complicated formula is explained in simple terms so that classification societies and ROs can help ship operators to implement the EEDI.

9. Resolution MEPC.324(75). (Adopted on 20 November 2020). *Amendments to the Annex of the Protocol of 1997.* https://wwwcdn.imo.org/localresources/en/OurWork/Environment/Documents/Air%20pollution/MEPC.324(75).pdf.

Regulation 21 of this resolution gives the reduction factors for the required EEDI relative to the EEDI reference line. A table is provided that outlines various ship types and tonnages as per the timeline. The last of these reduction factors is applicable from 1 January 2025 onwards.

10. Implementing Energy Efficiency Design Index. https://www.irclass.org/media/2368/energy-efficiency-design-index.pdf.

This document by IRClass explains the basics of EEDI and its calculations. It also explains the reference line and its applications.

11. IMO Resolution MEPC.350(78). (Adopted on 10 June 2022). *2022 Guidelines on the Method of Calculation of the Attained Energy Efficiency Existing Ship Index (EEXI).* https://wwwcdn.imo.org/localresources/en/KnowledgeCentre/IndexofIMOResolutions/MEPCDocuments/MEPC.350(78).pdf.

MEPC resolution 350(78) deals with the calculation of EEXI. It also explains the calculation of the conversion factor between fuel consumption and CO_2 emission.

12. IMO. (n.d.). *Improving the Energy Efficiency of Ships.* https://www.imo.org/en/OurWork/Environment/Pages/Improving%20the%20energy%20efficiency%20of%20ships.aspx.

This IMO document gives a general guideline on improving the energy efficiency of ships. It also states methods of improving the EEDI of new ships.

13. IMO Resolution MEPC.355(78). (Adopted on 10 June 2022). *2022 Interim Guidelines on Correction Factors and Voyage Adjustments for CII Calculations (CII Guidelines, G5).* https://wwwcdn.imo.org/localresources/en/KnowledgeCentre/IndexofIMOResolutions/MEPCDocuments/MEPC.355(78).pdf.

This resolution gives the attained annual operational CII for voyage adjustments and correction factors. It also deals with the correction factors related to electrical power on refrigerated containers as well as the cargo cooling systems on gas carriers and LNG carriers.

14. IMO MEPC Resolution 336(76). (Adopted on 17 June 2021). *2021 Guidelines on the Operational Carbon Intensity Indicators and the Calculation Methods.* Accessed on 31 March 2024. https://wwwcdn.imo.org/localresources/en/OurWork/Environment/Documents/Air%20pollution/MEPC.336(76).pdf.

The calculation methods for the operational CII are enumerated here.

Chapter 4

Strategies for Ship Energy Efficiency

Global warming is a serious problem globally, which can have devastating effects if not attended to urgently. The United Nations has urged all countries and member organisations to take stern measures to control GHG emissions from all sources.

The shipping industry, with a 3% share of global emissions, plays a critical role in mitigating global warming and achieving sustainability objectives through emissions reduction. The shipping industry in general, and ship-owners in particular, can contribute to a more sustainable future by adopting sustainable practices and complying with the regulatory requirements of International Maritime Organization (IMO).

IMO has brought about a slew of measures towards this. The aim is to reduce GHG emissions by half from current levels by 2050. It seems an achievable target provided the ship-owners and other stakeholders do their bit. In this chapter, we will discuss the IMO strategies for the reduction of GHG emissions from ships, with the primary goal of limiting global warming. We will also look at how these strategies can be implemented and what are the roles of the various stakeholders related to this.

4.1 IMO STRATEGIES FOR EMISSION REDUCTION

Globally, many initiatives and treaties have been developed in order to combat GHG emissions. As part of the United Nations' focus on limiting global warming, IMO introduced the initial strategy for GHG reduction in 2018 followed by the final strategy in 2023. We will now discuss these strategies and the requirements for complying with them.

4.1.1 Initial IMO Strategy

The 2014 Third IMO GHG Study projected that in 2012, GHG emissions from global shipping constituted approximately 2.2% of human-caused carbon dioxide emissions. The study further indicated that these emissions might increase by a range of 50% to 250% by the year 2050. This prompted the IMO to take quick action to curb GHG emissions from ships.

DOI: 10.1201/9781032702568-4

The initial strategy was specified in IMO res 304(72), which entered into force on 13 April 2018. It introduced three levels of ambition as follows:

- To reduce the carbon intensity of ships by implementing further phases of the EEDI for new ships, by incorporating percentage improvement for each phase depending on the ship type.
- To reduce the carbon intensity of international shipping by reducing the CO_2 emissions per transport work by at least 40% in 2030 and striving for 70% as compared to the 2008 levels.
- To reduce the GHG emissions by at least 50% by 2050 as compared to 2008 levels and strive to phase out CO_2 emissions altogether in line with the Paris Agreement temperature goals.

The resolution also set out short-term, mid-term, and long-term measures to combat GHG emissions from international shipping as per the initial strategy as follows:

- **Short-Term Measures:** These are those measures finalised and agreed by the committee between 2018 and 2023. These measures include the following:
  - Advancing the EEDI and the Ship Energy Efficiency Management Plan (SEEMP)
  - Formulating operational indicators for both new and existing vessels
  - Instituting an existing fleet improvement programme
  - Adopting speed optimisation and reduction strategies
  - Creating and revising national action plans
  - Strengthening technical cooperation initiatives overseen by the IMO.
  - Advancing port infrastructure (e.g., onshore power supply from renewable sources).
  - Providing incentives for pioneers in the adoption of new technologies, among other measures.
- **Mid-Term Measures:** These are to be finalised and agreed between 2023 and 2030 and include the following:
  - Enactment of a programme to facilitate the widespread adoption of alternative low-carbon and zero-carbon fuels
  - Measures for both new and existing ships for operational energy efficiency
  - Market-based measures to encourage reduction of GHGs
  - Encourage technical cooperation using an Integrated Technical Cooperation Plan (ITCP)
  - A mechanism to obtain feedback and thereby learn lessons on the effects of the implementation of these measures.
- **Long-Term Measures:** These include those that are to be finalised and agreed upon beyond 2030. They include endeavours like advancing the development and supply of zero-carbon or fossil-free fuels.

Additionally, efforts may focus on encouraging and facilitating alternative innovative mechanisms for reducing emissions.

The MEPC Resolution 304(72),[15] adopted on 13 May 2018, details the vision, the levels of ambition, and the guiding principles for the initial strategy. The resolution also sets out the short-term, mid-term, and long-term measures with possible timelines and their impact on the states.

4.1.2 Revised IMO Strategy

The third IMO GHG Study 2014 showed that in the year 2012, GHG emissions from global shipping amounted to 2.2% of the anthropogenic CO_2 emissions. This rose to almost 2.89% by 2018, that is, an increase of 30% between 2012 and 2018. Further studies showed that if immediate measures are not implemented, the emissions are expected to rise by between 50% and 250% by the year 2050.

This was not good news for the international community, and IMO decided to revise their levels of ambition and the strategy to combat GHG emissions from shipping. Against this backdrop, the MEPC held its 80th session (MEPC 80) from 3 to 7 July 2023 and brought out Resolution 377(80), known as the 2030 revised IMO strategy on the reduction of GHG emissions from ships. This resolution was adopted on 7 July 2023, and the 2018 initial IMO strategy was replaced by the 2023 revised strategy.

The 2023 IMO GHG strategy introduced four levels of ambition as follows:

- Energy efficiency of new ships to improve further so as to reduce the carbon intensity of the ship.
- CO_2 emissions per transport work to reduce by at least 40% by 2030 as compared to 2008 levels, resulting in a reduction in the carbon intensity of international shipping.
- Increase in zero or near-zero emission technologies, fuels, or energy sources so that they account for at least 5%, striving for 10% of the total energy used by global shipping.
- GHG emissions from global shipping to decline steadily so that it reaches net zero by or around 2050.

To achieve these ambitions, MEPC 80 has given further clarification by way of indicative checkpoints. These are targets which need to be achieved along the way so that net-zero emissions may be achieved by 2050. These indicative checkpoints are as follows:

- To achieve a minimum of a 20% reduction, with an aspiration for 30%, in the total annual GHG emissions from international shipping by 2030, relative to the levels recorded in 2008.

- To achieve a minimum reduction of 70%, with an aim for 80%, in the total annual GHG emissions from international shipping by 2040, relative to the levels observed in 2008.

The 2023 strategy also revised the short- and mid-term measures as follows:

- **Short-Term Measures:** To conduct a review of the technical and operational measures as per MARPOL Annex VI. This review is to be completed by 1 January 2026.
- **Mid-Term Measures:** These consist of a technical element consisting of a fuel standard for reducing GHG emissions and an economic element consisting of a pricing mechanism for GHG emissions.

The MEPC Resolution 377(80),[16] adopted on 7 July 2023, specifies the levels of ambition and the guiding principles, as well as the impact of these measures on the member states (IMO, 2023).

Now that we're aware of the IMO strategy, let's discuss what measures can be put in place to implement them. It is important to understand the IMO strategies so that they can be implemented by the ship-owners so that the emission of GHGs will be reduced, leading to the required zero-carbon shipping by 2050.

4.2 HOW CAN THESE STRATEGIES BE IMPLEMENTED?

The following are some of the measures that can be adopted by ship-owners and others in order to reduce their carbon footprint and become energy efficient in line with the IMO's GHG strategy:

- **Hull Design and Maintenance:** The design of a ship's hull plays a crucial role in minimising energy usage. The resistance caused by wind and water results in frictional losses, which in turn increase fuel consumption and diminish overall energy efficiency. Therefore, emphasis is placed on optimising the hydrodynamic characteristics of the hull. This can be done as follows:
  - Shaping the bow and stern, as well as refining the hull's contours, can substantially diminish frictional losses and consequently decrease energy consumption. At the design phase, efforts are made to streamline the hull in order to mitigate resistance and turbulence, all with the goal of enhancing energy efficiency by reducing frictional losses.
  - The bulbous bow minimises the formation of a bow wave which causes wave resistance. It ensures a streamlined flow of water along the hull, thereby reducing frictional resistance. This increases the fuel efficiency and thereby reduces fuel consumption, which results

in reducing the GHG emissions and is also cost-effective for the ship-owner.

- The shape of the stern can also increase fuel efficiency. It reduces the wetted surface area of the stern, leading to lesser drag and water resistance. This will reduce fuel consumption and result in emission of GHGs.
- The maintenance of the hull is another major factor for energy efficiency. Conventional antifouling paints can reduce the formation of sea growth on the hull but not prevent it. This results in barnacles and other growth on the hull, resulting in frictional losses leading to increased fuel consumption and lesser speed. Ship-owners and operators have to invest in superior paints like silicon-based paints which prevent the formation of marine growth. Further increased hull inspection and hull cleaning will result in a cleaner hull and subsequently better fuel efficiency.
- **Propulsion Systems:** Advancements in propulsion systems, such as more efficient engines and propulsion methods (e.g., LNG propulsion), can significantly reduce fuel consumption and emissions. This is because incorporating new technologies in marine engines can convert a higher percentage of the energy from the fuel into useful mechanical work, resulting in reduced fuel consumption and lower emissions. Modern engine designs incorporate features such as better combustion efficiency, reduced friction, and improved thermal management to maximise their efficiency.

  Propeller design plays a critical role in the efficiency of propulsion systems. With innovative technology, propellers can now be optimised for specific operating conditions, such as vessel speed, draft, and sea state, to maximise thrust while minimising energy losses due to cavitation and vibration. Variable pitch propellers enable the modification of propeller blade angles to enhance performance in response to changing load conditions. Through dynamic adjustments of pitch according to engine load and operational factors, these propellers have the capacity to enhance fuel efficiency and minimise engine wear and tear. These propellers have control on the bridge, which allows pitch variations as per requirements.

  Some smaller ships are being fitted with Air Lubrication Systems. They inject a layer of air bubbles beneath the hull. These help to reduce frictional losses and thereby increase fuel efficiency.
- **Energy Management Systems:** Implementing EMS on ships allows for real-time monitoring and optimisation of energy consumption. These systems can control lighting, air conditioning, and other on-board systems to reduce unnecessary energy use. EMS can also automate load shedding, which involves selectively cutting power to non-essential equipment or systems to optimise power distribution and reduce energy consumption when necessary. EMS for ships continuously

monitors and collects data from various sensors and instruments placed throughout the vessel. This data includes information on engine performance, fuel consumption, power demand, weather conditions, and other relevant parameters. It can optimise engine load, speed, and fuel injection to achieve the best balance between power requirements and fuel consumption.

In general, EMS is important to curb fuel consumption and improve the overall energy efficiency aboard ships. Real-time monitoring, optimised operation, load management, predictive maintenance, integration with weather routing, and performance tracking are some of the measures that EMS incorporates to increase energy efficiency. EMS empowers ship operators to mitigate energy wastage, slash operating expenses, and diminish environmental footprint.

- **Hybrid and Electric Systems:** Hybrid and electric propulsion systems, combined with energy storage solutions like batteries, can be employed to minimise fuel consumption during low-power operations or when manoeuvring in ports. These systems combine electric propulsion with traditional diesel engines to increase fuel efficiency. Further, they can also be used with renewable energy sources and switch from traditional fuels to renewable energy depending on the situation. They rely on advanced energy management and control systems to determine when to switch between the internal combustion engine and electric motor to operate the vessel in the most energy-efficient manner.

  Electrical propulsion comes with its own challenges regarding costs, infrastructure, performance, and range. Further, any electrical system will reduce GHG emissions only if it is sourced from non-fossil fuels. Currently, electricity generation depends largely on fossil fuels. Hence, the environmental impact of such systems is reduced somewhat. But power plants are increasingly reducing GHG emissions using advanced technology and streamlining their processes. The full potential of electric and hybrid systems will be realised when these challenges are overcome.

- **Waste Heat Recovery:** Ships generate a substantial amount of waste heat during their normal operations, which can be harnessed and converted into useful energy, thereby reducing fuel consumption and GHG emissions. Waste heat on ships can come from various sources, including engine exhaust gases, exhaust gas boilers, the main engine cooling system, auxiliary engines, and even cargo heating systems. WHR systems can be used for electricity generation, heating and cooling of spaces such as cabins and cargo holds, as well as desalination and freshwater production. Waste heat recovery is an important strategy to increase fuel efficiency on board ships. But for older ships, newer technology may be needed, and retrofitting these may be a complicated and expensive process.

These systems enhance energy efficiency on-board ships by capturing and utilising waste heat generated during combustion of the fossil fuels. By recirculating waste energy for heating, power generation, steam production, and hybrid propulsion, WHR systems help reduce fuel consumption, lower emissions, and contribute to sustainable shipping.

- **Voyage Planning:** By optimising routes and making decisions based on real-time weather and oceanographic data, vessels can reduce fuel consumption and GHG emissions. Adjusting speed to reduce resistance from rough seas or to take advantage of favourable weather conditions contributes to energy efficiency. Weather routing services prioritise the safety of the vessel, its crew, and the cargo. Environmental concerns, such as reducing emissions and protecting sensitive marine ecosystems, are also factored into the routing decisions.

Voyage planning is an important factor in improving energy efficiency. Optimisation of the route and speed, efficient cargo stowage, and smart navigation strategies all help to reduce fuel consumption and GHG emissions. By incorporating factors such as weather conditions, traffic patterns, and environmental regulations into the planning process, ships can minimise fuel consumption, reduce emissions, and enhance overall operational efficiency during voyages.

- **Slow Steaming:** Operating at reduced speeds, known as slow steaming, is an effective method to conserve fuel, although it may increase voyage duration. Slow steaming results in reduced engine output and decreased fuel consumption, leading to a decrease in the release of GHGs and air pollutants into the environment. The practice of slow steaming aligns with the shipping sector's objectives to achieve emissions reduction goals and adhere to environmental guidelines. Nevertheless, shipping firms must meticulously manage the trade-off between fuel conservation and meeting delivery deadlines and customer requirements. Slow steaming might not be the optimal choice for shipments with stringent time constraints.

Slow steaming is not without its own disadvantages. Most importantly, it increases the voyage duration, which may not be acceptable to charterers and shippers. Further, slow steaming leads to maintenance issues, such as soot collection. The carbon and soot generated are not thoroughly flushed out due to lesser exhaust gas pressure during slow streaming. The ship's crew has to make speed adjustments regularly to clear the soot. This process is known as soot blowing, without which there is a real fear of a fire in the scavenge space of the engines. The ship master and operator must take careful consideration of these factors before proceeding at slow speeds.

- **Energy-Efficient Technologies:** Energy-efficient solutions on ships play a vital role in reducing fuel consumption, operational expenses, and the discharge of GHGs. The installation of energy-efficient devices and systems on board, such as LED illumination and HVAC systems designed for energy efficiency, can result in significant reductions in energy consumption. Forced induction technologies, such as turbochargers and superchargers, increase engine efficiency by compressing air to provide more oxygen for combustion. There are other options available which we shall discuss in our section on intelligent engines. The installation of emissions control systems, such as scrubbers and selective catalytic reduction (SCR) systems, can help reduce air pollutants from exhaust gases.

- **Alternative Fuels:** The use of alternative fuels in ships has gained increasing attention in recent years as the maritime industry seeks to reduce its environmental impact and comply with stringent emissions regulations. Exploring alternative fuels, like biofuels or hydrogen, can be a sustainable way to reduce emissions and reliance on traditional fossil fuels. We shall discuss the alternate fuels and their uses in the section on low carbon operations.

- **Crew Training:** Proper training and awareness among the crew regarding energy-efficient practices and the use of on-board systems can contribute to better energy management. They should receive training in the best practices, including proper engine operation, load management, and fuel-efficient navigation techniques. Currently, there is no mandatory requirement for crew training in energy efficiency. Shipowners and operators must consider in-house training for their crew so that they understand the importance of reducing fuel consumption and thereby GHG emissions. On board ships, senior officers must explain the contents of SEEMP to them and motivate them to follow its requirements. With new technologies being incorporated on the ship to reduce fuel consumption, it has become necessary for the ship's crew to receive training on the use and maintenance of such energy-efficient technologies and equipment. To ensure that crew training is carried out in a systematic manner, IMO has developed a Train the Trainers course on energy efficiency. This will be discussed later in this chapter.

- **Regulatory Compliance:** Compliance with international regulations, such as the IMO's EEXI and the EEOI, is essential for reducing emissions and improving energy efficiency. Port state control officers routinely visit ships in port to confirm whether they are in compliance with the energy efficiency measures as specified by IMO and various conventions in force.

The best practices for ensuring energy efficiency on board ships are discussed in later chapters. However, it is important to note that improving energy efficiency in ship operations isn't solely the responsibility of individual ship management; it involves various stakeholders as well. Let's discuss this in the next section.

4.3 UNDERSTANDING THE ROLES OF VARIOUS STAKEHOLDERS

Ships do not work in isolation; rather, shipping is a community. All the stakeholders have a role to play in the reduction of GHG emissions and thereby control global warming. It is therefore important to know who the stakeholders are and what their role is in energy efficiency. The following are the key stakeholders:

- **Industry Organisations**: IMO, BIMCO, and other organisations provide guidance, best practices, and support for the implementation of SEEMP. The Baltic and International Maritime Council provides knowledge and advice to their members on shipping-related matters. They also offer resources and training to help ship-owners and operators enhance energy efficiency. Their efforts in facilitating communication and collaboration between the various stakeholders in the maritime industry for the sake of increasing energy efficiency are paramount if the IMO's ambitions as spelled out in the revised GHG strategies are to be met. In addition to BIMCO, International Chamber of Shipping (ICS), International Association of Independent Tanker Owners (INTERTANKO), and others also provide guidance on reducing GHG emissions from ships.

  Industry organisations contribute to a more sustainable future by knowledge sharing, issuing best practices, providing education and training for ship and shore staff, and helping in policymaking, as well as through research and development in the field of reducing emissions from the ship.
- **Classification Societies:** These are non-governmental bodies that establish and maintain technical standards for the construction and operation of ships. They ensure that ships meet certain safety and environmental standards. They play a key role in verifying and certifying SEEMPs, as well as providing guidance on energy-efficient practices. They look after the ships' design features right from the construction stage and therefore are entrusted with the planning, implementation, review, and approval of the vessels' energy-efficient systems and equipment. They are often entrusted with conducting surveys and audits related to energy efficiency and issuing relevant certificates on behalf of the flag administration. They publish newsletters and circulars advising ship-owners on the latest developments in energy efficiency and help the ships comply with the requirements of energy efficiency.

The work of Classification Societies is pivotal in promoting energy efficiency on ships to ensure the reduction of fuel consumption and subsequently reduce GHG emissions.

Classification Societies are key players in promoting sustainability in shipping. They develop and enforce standards for ship design and construction aligned with the IMO requirements for energy efficiency and GHG reduction. By providing technical support, training, and education for ships' crew, as well as conducting research and publishing guidelines on sustainable practices, they assist the ship operators in ensuring that their ships are in compliance with IMO GHG strategies and reduce their environmental impact.

- **Port Authorities:** They contribute to energy efficiency by providing timely and accurate information to ships. This includes information on pilot boarding, berth availability, and weather conditions, which enable ships to optimise their arrival schedules and reduce the need for excessive speed or anchoring. This "just in time" approach can significantly save energy and reduce emissions. They invest in energy-efficient infrastructure to reduce emissions. Ports also provide the facility of cold ironing (providing shore power to the ships) so that ships can shut down their auxiliary engines and thereby reduce GHG emissions from ships while in port. Further many ports provide incentives and initiatives to ships to encourage ship-owners to reduce emissions from their ships.

  Ports play a role in promoting sustainability by implementing green port initiatives, implementing environmental management practices, and adopting sustainable practices in port operations. Further, they ensure that the ships visiting their ports are in compliance with the GHG strategies and other regulations of MARPOL.

- **Charterers and Customers:** They can influence energy efficiency by demanding information on a ship's fuel efficiency and environmental performance as part of their selection criteria. They can also incentivise ship-owners and operators to adopt more eco-friendly practices by offering favourable terms or contracts based on the ship's environmental performance. They can also optimise the vessel's voyage and speed in order to increase fuel efficiency. In addition, they have the power to select those ships that are energy efficient and thereby apply commercial pressure on the ship-owners.

  Charterers and other customers enjoy purchasing power over the ships they employ. They can thus drive the demand for greener technologies and engage in stakeholder collaboration to reduce the harmful emissions from ships. By efficient cargo handling processes, improved supply chain management, and leveraging their influence over the ship operators, they can contribute to reducing global warming in general.

- **Insurers:** Insurers can also put pressure on the ship-owners to reduce their GHG emissions by allowing a discount on the premium for those ships that have incorporated advanced technologies in order to increase their energy efficiency. Further, vessels that are operating on alternative fuels or renewable sources of energy can also be given a discount on the insurance premium. Ship-owners have to take two types of insurance:

- A hull and machinery insurance. This covers the ship and her equipment.
- An insurance for third-party damage, usually from Protection and Indemnity Clubs (P & I Clubs).
- If both these insurers offer a better premium for compliant vessels, then the ship-owners will ensure compliance well ahead of the target dates. In addition, many insurers are adopting the Poseidon Principles to ensure that ships are in compliance with the IMO GHG strategies. We shall discuss this in detail later in Chapter 9.

In conclusion, we can say that it is not only the responsibility of the ship and the ship-owner to increase fuel efficiency. All stakeholders have a role to play to protect the environment and reduce global warming. It is only when all come together that the vision of IMO to achieve net zero and thereby prevent global warming can be successful.

In their pursuit to spread awareness, the IMO has introduced the Train the Trainer (TTT) course on energy-efficient ship operation. This course is intended to train the trainers who will impart the training to the ship staff and others. Let's look at this in detail.

4.4 TRAIN THE TRAINER COURSE ON ENERGY-EFFICIENT SHIP OPERATIONS

Energy efficiency is a relatively new topic that requires some knowledge on the part of the ship staff in order to comply with the requirements. Thus, IMO recommends that personnel be trained in the various aspects of energy efficiency. The trainers, therefore, have to be trained in the subject if they are to do justice to the training. Keeping this in mind, IMO developed the TTT course on energy-efficient ship operations.[17]

This course focuses on the management aspects and operational aspects of limiting GHG emissions from ships. It provides emission solutions and an EEDI calculator for training purposes. It comprises six study modules:

- Module 1: Climate change and shipping response
    This module deals with the impact of GHG emissions on climate change, as well as the efforts of the international community and IMO to combat this through legislation, technical cooperation, and transfer of technology.

- Module 2: IMO energy efficiency regulations and related guidelines
  This module deals with the regulatory requirements, specifically MARPOL Annex VI. It also explains the attained EEDI calculations, the SEEMP development guidelines, as well as the EEOI calculations.
- Module 3: From management to operation
  Module 3 deals with the ship operations affecting energy efficiency, such as fleet optimisation, slow steaming, just-in-time arrival, weather routing, ship capacity utilisation, and other operational factors that can help to reduce fuel consumption.
- Module 4: Shipboard energy management
  This module covers one of the most important aspects of on-board energy efficiency. It deals with the shipboard roles and responsibilities, and the management of the ship's trim and ballast water in order to increase energy efficiency. Engine management, maintenance of hull and propellers, and other ship-related factors are also included.
- Module 5: Ship-port interface and energy efficiency
  Ports are an integral part of ship operations. Therefore, ship-port interface, port operation and management, setting up green ports, programmes related to the environment, and review of energy management are included in this module.
- Module 6: Energy management plans and systems
  This module contains the review of on-board energy management plans and systems in place for their implementation. ISO 50001 is a quality system for energy management and is included in this module. It also contains details of energy planning, energy audits, and company energy management systems, as well as data collection, monitoring, and reporting.

In conclusion, the TTT course on energy-efficient ship operations deals with the various aspects of energy efficiency. Trainers imparting energy efficiency training to both ship and shore staff are expected to complete this course.

Training is an important aspect of promoting sustainable operations on ships. Investing in training and education has multiple benefits for the organisation and will ensure a positive change in the mindset of their staff, leading to increased awareness and compliance with regulations.

4.5 SUMMARY

In this chapter, we have discussed IMO's initial and revised strategy for GHG reduction and how they are relevant to the ship. We have seen what steps the ship-owners and other stakeholders can take to implement this strategy and work towards the reduction of GHGs. Training is an important aspect of any meaningful measures on board, and with this in mind, IMO has designed the TTT course for energy-efficient ship operations. Overall, this chapter provides an understanding of IMO's strategies and ways to implement them.

In the next chapter, we shall discuss how monitoring of fuel consumption and machinery performance can help increase fuel efficiency.

BIBLIOGRAPHY

15. MEPC Res 304(72). (Adopted on 13 May 2018). *Initial IMO Strategy on Reduction of GHG Emissions from Ships.* https://wwwcdn.imo.org/local-resources/en/KnowledgeCentre/IndexofIMOResolutions/MEPCDocuments/MEPC.304(72).pdf.

The initial IMO strategy is outlined in the resolution which has since been superseded by the 2023 final strategy. The resolution states the levels of ambition and the guiding principles in order to reduce GHG emissions from ships.

16. IMO Res 377(80). (2023). *IMO Strategy on Reduction of GHG Emissions from Ships.* https://wwwcdn.imo.org/localresources/en/MediaCentre/Press-Briefings/Documents/Resolution%20MEPC.377(80).pdf.

This resolution, commonly known as MEPC 80, is the foundation stone for IMO to enforce the new strategy which seeks to ensure zero emissions by 2050.

17. International Maritime Organization (IMO). *IMO Train the Trainer (TTT) Course on Energy-Efficient Ship Operation.* https://www.imo.org/en/Our-Work/Environment/Pages/IMO-Train-the-Trainer-Course.aspx.

This IMO document details the complete package for conducting the course on energy-efficient operations. It includes the trainer's manual and the various course modules to ensure that the course syllabus is uniform across the world.

Chapter 5

Monitoring Fuel Consumption

Fuel efficiency and reducing the consumption of fossil fuels are the best ways to control the emission of greenhouse gases (GHGs) and global warming. Monitoring fuel consumption is therefore of paramount importance if we are to seriously increase fuel efficiency. Monitoring is the first step towards identifying whether the fuel consumption is higher than the norm. Only then can action be taken to reduce the same and thereby reduce the emission of GHGs.

In this chapter, we will discuss fuel monitoring systems and various methods of monitoring fuel consumption. Implementing engine monitoring systems is an effective way to monitor and reduce fuel consumption, and this will also be discussed. In addition, we will highlight the roles of the various officers on board who monitor fuel consumption and thereby take action to reduce it.

Let's start with a brief introduction to what exactly a fuel monitoring system is.

5.1 UNDERSTANDING FUEL MONITORING SYSTEMS

The best method to monitor fuel consumption is by implementing a fuel monitoring system. These are required for accurately measuring fuel consumption and managing the same. A fuel monitoring system would consist of the following components:

- **Flow Meters:** These are the most common and useful equipment to measure fuel consumption. Many types of flow meters are available such as positive displacement meters, electromagnetic flow meters, and turbine flow meters.
- **Fuel Density and Temperature Sensors:** These sensors are required to measure fuel consumption accurately. This is because the fuel properties can change with temperature and density and thereby compensation can be made for variations due to this.
- **Engine Performance Analysis:** Monitoring engine performance is crucial to reducing fuel consumption. An analysis of engine performance

 DOI: 10.1201/9781032702568-5

with follow-up action, where required, is one of the most important measures for fuel efficiency. We shall deal with this later in this chapter.

- **Alarms and Alerts:** These are a necessary requirement of any engine monitoring system. There are different alarms that provide indications of engine performance. They can be set up to alert the staff when fuel consumption exceeds predetermined levels or when other abnormalities are detected.
- **Communication and Connectivity:** In today's world, monitoring of engine performance can be done in real time by vessel superintendents sitting in the shore office. Due to satellite connectivity, they can study the performance of the engines and advise the chief engineer on board accordingly. In case of any serious issues, the manufacturers' specialists can be contacted for further advice. Thus, communication and connectivity are important aspects of engine performance.
- **Training:** As in any monitoring system, crew training is important if the monitoring is to be successful. The crew should be aware of the parameters to look out for and the immediate actions required. They should also be made aware of the best practices for energy efficiency.

The preceding points are pointers for monitoring and reduction of fuel consumption. It is for the shipping company to constantly monitor the fuel efficiency of their ships. Masters and chief engineers, on their part, must take the necessary steps to ensure that the machinery runs at optimum efficiency. They must also maintain proper records as evidence of energy efficiency so that regulatory compliance can be demonstrated to the administrations and their agencies.

Fuel monitoring systems help ship operators to optimise fuel usage, improve operational efficiency, and reduce energy consumption, thereby contributing to cost savings, environmental sustainability, and regulatory compliance in maritime operations.

Regulations are important for fuel monitoring systems as they establish standards, so that there is accuracy and reliability in the reporting process. They ensure that fuel monitoring systems operate efficiently, contributing to sustainable practices.

5.2 REGULATORY REQUIREMENTS FOR REPORTING FUEL CONSUMPTION

It has long been established that the amount of carbon dioxide emissions is directly related to fuel consumption. Measurement of EEDI and EEXI depends to a large extent on the fuel consumption of the ship. International Maritime Organization (IMO) has introduced mandatory reporting requirements for fuel oil consumption. Resolution MEPC 278(70), adopted on 28 October 2016 and introduced as an amendment to MARPOL Annex VI, requires all vessels to collect and report consumption data for each type of fuel used on board, starting from 1 January 2019. This information will be used to monitor fuel consumption and introduce measures to reduce it.

The data of fuel consumption of the ship is to be compiled and forwarded to the administration or the recognised organisation authorised by them. Once it is confirmed that the data has been reported as per the requirements of Annex VI, a Statement of Compliance will be issued to the ship related to the fuel oil consumption.

This system of reporting fuel consumption, known as the IMO DCS,[18] is instrumental in ensuring compliance of the ship-owners and the ships with the IMO GHG strategy.

Let us now discuss how monitoring of fuel consumption can lead to the next step of reducing fuel consumption.

5.3 METHODS FOR MONITORING AND REDUCING FUEL CONSUMPTION

There are some important on-board procedures whereby the engineers can track fuel consumption and take action to reduce it. For example, we know that well-maintained equipment is more efficient and gives a better performance than those which are not maintained well. On the ship, good maintenance of the engines will increase their efficiency and thereby result in reducing fuel consumption and promoting sustainable operations. Let us have a look at some measures that can be taken to help achieve this.

5.3.1 Hull-Propeller Performance

Hull propeller performance is the measure of the effectiveness of the propeller in converting engine power into thrust to move the vessel through water. Analysing the hull-propeller performance, specifically comparing the current engine power-speed curve to speed trials, is an essential step in assessing a ship's efficiency and identifying opportunities for improvement. Here are some common measures to assess and ensure hull-propeller performance:

- **Engine Power-Speed Curve:** The engine power-speed curve provides details about the performance of the engine and power output variations with the speed of the vessel. It represents the relationship between the ship's engine power (in kilowatts or horsepower) and its speed (in knots or metres per second). This curve is typically created during sea trials or through engine testing and is unique to each vessel. This graph is created as follows:
  1 The vessel's speed is recorded at various engine settings, and the power output is noted.

2 This data is compiled during the sea trials in order to determine the engine performance under various conditions.

3 The Engine power–speed curve is plotted with the engine speed on the X-axis and the power output on the Y-axis.

The curve will be non-linear, starting from zero at idling of the engines and increasing to maximum rated power.

The engine power-speed curve prepared at the time of sea trials provides a useful reference for estimating the engine performance during the operational life of the ship. The power-speed curve can be recreated at any time under ideal conditions and compared with the original curve to detect shortfalls in engine performance. These deviations can then be corrected on board if possible. If not, the engine manufacturers look into these deviations to come up with possible solutions.

Engine speed curves serve as a valuable tool for optimising energy efficiency by providing insights into the relationship between engine speed and fuel consumption. By leveraging the information provided by engine speed curves, ship operators can make informed decisions to maximise energy efficiency, reduce fuel consumption, and minimise environmental impact.

- **Machinery Health**: Fuel consumption is closely related to the health and condition of the ship's machinery, including the engine, propulsion systems, and auxiliary equipment. We have all seen how old and unmaintained vehicles can spew smoke, while a newer vehicle does not. Maintaining the health of ship machinery is crucial for optimising energy efficiency, reducing fuel consumption, minimising emissions, and enhancing overall operational reliability and safety.

- **Routine Maintenance**: Regular maintenance is the only way to keep machinery in good health. Every vessel has an approved planned maintenance system which helps the ship's engineers carry out routing inspections, overhauling, and maintenance as per the manufacturer's instructions and timelines. This helps prevent unexpected breakdowns and ensures that the machinery operates optimally. In addition, there are several checks and analyses that are carried out to determine the health of the engines so that timely corrective action can be taken in order to enhance the performance of the machinery and ensure fuel efficiency.

- **Vibration Analysis**: Vibration analysis is used to monitor the health of machinery and helps in increasing overall performance and fuel efficiency. Vibration data is collected from the engine by means of sensors placed at various locations. Based on this data vibration analysis is carried out, which involves measuring the amplitude, frequency, and phase of the vibrations. From the analysis, irregularities can be identified, such as worn-out bearings, wear and tear of moving parts, misalignment, unbalanced components, and so on. By this method, excessive vibrations can be determined, and their root cause identified.

Once the root cause has been identified, corrective action can be taken such as replacing worn-out parts and bearings, dynamic balancing of rotating components, re-alignment, and so on. Vibration analysis helps in enhancing the energy efficiency of ship engines by identifying mechanical issues, optimising performance, preventing failures, and taking corrective action before the breakdown occurs.

- **Resonance Analysis:** This is a method of identifying excessive vibrations. Every structure has its own natural frequency, and resonance occurs when the vibration is amplified. It involves assessing the dynamic response of the engine and its components to various operating conditions, including vibration frequencies and amplitudes. This can lead to excessive vibrations at the critical speed. This occurs at the critical RPM and leads to inefficiencies in the engines, including high torsional stresses, which can cause structural damage to the engines. Thus, avoiding the critical speed is important for the health of the ships' engines. Resonance analysis helps identify and address sources of energy loss due to mechanical inefficiencies, such as friction, imbalance, or misalignment. By optimising system performance, energy losses can be minimised, leading to improved overall energy efficiency and a reduction in fuel consumption.

- **Predictive Maintenance:** We have seen earlier in this chapter how planned maintenance can ensure that engines and other equipment can be properly maintained for optimum performance. Predictive maintenance is one step ahead that uses data, sensors, and analytics to predict the maintenance required. Machine learning algorithms collect data and predict issues at the earliest stage. This will help to take corrective action before time and prevent breakdowns, inefficiencies, and so on. Predictive maintenance can improve energy efficiency by minimising downtime, preventing energy waste, optimising equipment performance, and facilitating proactive maintenance strategies.

- **Condition Monitoring Systems:** This is another requirement of any engine management system to analyse the engine performance. Condition monitoring is required for routine maintenance as well as for predictive maintenance. Sensors collect data on various parameters, such as fuel consumption, engine speed and load, exhaust gas temperatures, vibration and noise, oil temperature and quality, and so on. Various engine manufacturers have different methods of collecting and analysing data. Based on this data, the scheduled maintenance, as well as predictive maintenance, is decided and carried out.

As an example, let's assume that the bearings of the fire pump are wearing out. The pump is still working fine, but its efficiency has gone down. It is not possible for the ship's engineers to understand this unless visible and audible signs appear. But the sensors pick up the slight vibration, the increase in the temperature, as well as the noise levels. Based on these, it is established that the bearings need to be changed. The chief engineer is informed accordingly and he opens up the fire pump to find that the bearings are indeed worn out. This analysis and predictive maintenance not only increase the efficiency of the pump but enhance the safety of the ship and prevent a possible delay or detention of the ship during port state control inspections.

Condition monitoring will result in improving energy efficiency by enabling early detection of performance degradation, optimising equipment performance, preventing unplanned downtime, reducing maintenance costs, enhancing system reliability, and facilitating data-driven decision-making processes.

- **Oil Analysis:** This is another tool for ensuring the performance and fuel efficiency of the ship's engines. Once the oil sample is analysed, anomalies, if any, can be studied, and action taken to prevent damage and inefficiencies in the engines. This analysis can also indicate whether an oil change is required. The record of routine oil analysis can be studied for trends and long-term tracking of the engine performance. Oil analysis should be regularly carried out by collecting the oil samples and sending them to the approved laboratories for testing as follows:

  1 **Water Content Analysis:** This helps check for water in the sample. This may be due to condensation, leaking seals and gaskets, cooling system leaks, seawater ingress, lack of maintenance, or improper maintenance, and so on. Once it has been established that there is water in the oil sample, the reason can be determined. If corrective action is not taken immediately, corrosion, increased wear and tear of components, as well as insufficient lubrication, may take place. All these lead to deteriorating efficiency of the engines and increased fuel consumption.

  2 **Oil Analysis:** This detects levels of metals due to wear and tear. If undetected and uncorrected, these can cause excessive wear and tear of the engine components, leading to inefficiencies.

  Oil analysis serves as a proactive maintenance tool that contributes to energy efficiency by maintaining lubricant quality, detecting and addressing issues early, optimising lubrication practices, and extending equipment life. By leveraging the insights provided by oil analysis, organisations can minimise energy consumption, reduce operating costs, and improve overall productivity.

5.3.2 Engine Monitoring Systems

Many ship-owners prefer to install an engine monitoring system (EMS) on board. The EMS helps to increase fuel efficiency on ships by data crunching and providing real-time insights into the engine performance. Instead of piecemeal data analysis, the EMS gives a holistic picture and enables the ship staff and operators to operate the engine at full efficiency. This will result in the reduction of fuel consumption with associated reduction of

GHG emissions and cost benefits for the ship-owner. It consists of several inputs as follows:

- **Control Units and Software:** These are the heart and soul of the EMS. It consists of a central processing unit (CPU), which processes the incoming data. It can be considered the brain of the EMS and serves as an interface for human-machine interaction (HMI). The CPU is the hardware that collects all the data and forwards it to the software for processing. The software then gives instructions to the CPU in case any anomalies are detected and action is required. For example, if the carbon dioxide emission is found to be higher than normal, incomplete combustion may be the reason. The CPU passes on this info to the software, which analyses the data. It contains algorithms for corrective action, such as optimising fuel injection, managing exhaust systems, adjusting air intake, and so on. These instructions are passed on to the CPU, which carries out these functions.
- **Engine Overload Alarm:** Triggers when the engine is operating beyond its designed capacity. Signals the need to reduce engine load or adjust the parameters.
  - **High Exhaust Gas Temperature (EGT) Alarm:** Indicates inefficient combustion or engine stress. Corrective action may include adjusting the air-fuel ratio, checking for worn components, reducing engine load, and so on.
  - **High Exhaust Emission Alarm:** Sounds when excessive exhaust gas emissions are detected, such as NO_x and SO_x. Signals the need to reduce emissions by adjusting engine operations, improve combustion and thereby reduce the emissions.
  There are many other alarms, such as those indicating low fuel level, abnormal vibrations, fuel pump malfunctions, low lubricating oil pressure, and so on. Each of these alarms helps to identify certain issues, which if corrected, will increase engine efficiency, thereby helping to reduce fuel consumption and related GHG emissions.
- **Sensors:** These play an important role in the EMS. They include the following:
  - Speed sensors detect shaft speed and engine RPM.
  - Vibration sensors monitor the vibration of engines and other machinery.
  - Temperature sensors monitor the temperature of exhaust gas, cylinder heads, oil, engine coolant, and so on.
  - Pressure sensors measure oil and fuel pressures as well as cylinder and air bottle pressures.
  - Level sensors keep tabs on the fluid levels in various systems, such as fuel and oil tanks, and raise the alarm when the level falls below predetermined levels.

- **Engine Parameters:** These parameters are monitored in real time to identify any abnormalities. They include the load on the engines, fuel consumption, cylinder condition, and temperatures. Coolant and lubrication pressures and systems, as well as exhaust gas emissions, are monitored.
- **Communication Systems:** These are the essence of any monitoring system. Satellite connectivity enables real-time transmission of data from ship to shore for effective analysis. This allows for remote monitoring and enhances the effectivity of the EMS.
- **Equipment and System Manufacturers:** They are an important part of the EMS as they provide technical support and assistance. They can be contacted for analysing and interpreting data in case of major abnormalities. Further, they can contribute to fuel efficiency in the following ways:
  - Design and development are the forte of engine manufacturers. They constantly try to innovate on the engines using the latest technology in order to fine-tune the engines for fuel efficiency.
  - Fuel management systems are designed and operated by manufacturers on a real-time basis so that ship operators can monitor and control fuel consumption.
  - Energy-efficient technologies are offered by the manufacturers, which ship-owners can opt to incorporate on their ships, thereby aiding fuel efficiency.
  - Anti-fouling coatings that deter any sort of growth are available. This will ensure that the hull remains smooth and frictional losses are reduced.
  - Renewable energy solutions are available, where integration between conventional fuel and renewable sources helps the ship to automatically supplement conventional fuel in order to increase fuel efficiency.
  - Shipbuilders and manufacturers use advanced and lightweight materials in ship construction in order to reduce the weight of the ship and equipment, thereby enhancing fuel efficiency.

EMS is essential in the efforts to optimise the efficiency of the ship's engines and thereby reduce fuel consumption and exhaust gas emissions. It is essential that the ship and shore staff are aware of the potential of the EMS and work in tandem so that the EMS becomes an effective tool.

EMS plays a critical role in improving energy efficiency on ships by facilitating proactive maintenance and ensuring compliance with environmental regulations. They provide valuable insights and data that enable ship operators to optimise engine performance, reduce fuel consumption, and minimise environmental impact, ultimately leading to improved energy efficiency.

The European Union has published guidelines and best practices on monitoring and reporting fuel consumption, CO_2 emissions, and other relevant parameters.[19] This gives a comprehensive methodology for monitoring fuel consumption from the voyage planning stage. Now let us have a look at the most important cog in the wheel of the EMS, mainly the personnel who constantly monitor the engines and take corrective actions to ensure a reduction in fuel consumption:

- **Master:** As the overall in-charge of the ship, the master is responsible for implementing Ship Energy Efficiency Management Plan (SEEMP). It is the job of the master to ensure that the crew follows the best practices for energy efficiency as detailed in the SEEMP. They have to take an active interest to ensure that the vessel does not violate any provisions of MARPOL and other conventions. They have to maintain proper records and report to the shore office and the Administration as required. The master of the ship plays a pivotal role in driving energy efficiency initiatives, ensuring compliance with regulations, and promoting a culture of conservation on board. Their leadership and decision-making influence the vessel's fuel consumption as well as the environmental footprint and operational efficiency. The master plays a role in ensuring that the engine monitoring system is effectively implemented to improve energy efficiency and operational performance on board the ship. Their leadership, decision-making, and coordination efforts are essential for achieving fuel savings, reducing emissions, and enhancing sustainability in maritime operations.

- **Navigational Officers:** The chief officers and other navigational officers are responsible for watchkeeping. They have to ensure compliance with the vessel's SEEMP, including voyage and speed optimisation. They have to analyse EMS data related to fuel consumption, speed, and so on, and maintain reports and documents. They also have to coordinate with the engineering department to ensure compliance with MARPOL Annex VI regarding energy efficiency.

The EMS is not isolated in the engine alone but encompasses the entire ship's operations. The navigational officers work with the engineering team to ensure alignment between navigation practices and engine performance. By coordinating efforts, they optimise the ship's overall energy efficiency and operational effectiveness. For example, cooperation between the nautical and engineering departments is essential during deballasting to ensure that the ballast tanks have been fully emptied. Miscommunication between them will lead to um-pumpable ballast in the tanks, which will result in increased deadweight of the ship and affect the energy efficiency of the ship. Their coordination and cooperation ensure that operational practices align with fuel efficiency goals while maintaining safety and operational integrity.

- **Chief Engineer:** The chief engineer is responsible for the ship's engine room and all related operations. They have to follow the SEEMP and ensure it is implemented on board. They are responsible for monitoring the EMS, analysing EMS data, and ensuring that maintenance, repairs, and so on, are carried out as per the EMS. They have to ensure that the vessel is in compliance with MARPOL Annex VI, especially regarding exhaust gas emissions. They have to report to the shore office with all data as per the DCP and also report any abnormalities. The chief engineer has to liaise with the shore office for any resources or if the shore workshop assistance is required. Specifically, the chief engineer of the ship is responsible for the following actions in order to improve energy efficiency:
  - **Engine Performance:** The chief engineer's responsibility is to ensure that the main engines and auxiliary machinery are maintained well and operate at their highest efficiency levels, fine-tuning fuel combustion, and minimising energy wastage. These efforts directly translate into decreased fuel usage and emissions.
  - **Efficient Use of Resources:** The chief engineer oversees the allocation of the ship's energy resources, encompassing fuel, lubricants, and electricity. They enact strategies to curtail energy wastage, including optimising load distribution, reducing idle periods, and integrating energy-conserving technologies such as WHR systems or variable frequency drive.

  Thus, we see that the chief engineer holds a central position in the engine monitoring system by endeavouring to ensure that the ship's engines and machinery work at optimum levels at all times. These endeavours result in decreased fuel consumption, reduced operational expenses, and diminished environmental repercussions in maritime operations.

- **Second Engineer and Engineering Staff:** They work with the chief engineer and follow instructions regarding the engine room operations. They are responsible for monitoring the alarms and alerts and taking corrective actions where required. They assist the chief engineer with monitoring the EMS and conduct maintenance work on the engines. In case of any serious shortcomings, they have to report to the chief engineer immediately.

  The second engineer and engineering staff are the main actors in the success of the engine monitoring systems to reduce fuel consumption and enhance energy efficiency aboard ships. Their continuous efforts ensure a reduction in fuel consumption, operational efficiency, and environmental sustainability.

- **Company Superintendents and Marine Managers:** They are the link between the ship and the shore office and crucial in the success of the engine monitoring system. They are responsible for ensuring that the ship follows the requirements of SEEMP and energy efficiency. Many companies obtain real-time data from the ship through satellite connectivity. Others depend on the ship's reports regarding consumption data. In any case, the companies provide assistance and support to the ship to comply with the requirements. They monitor the EMS data and guide the chief engineer in case any adjustments, repairs, and so on, are required. They develop company policies and procedures aimed at improving energy efficiency and reducing fuel consumption across the fleet. These policies are implemented on board by the ship staff with the assistance of the shore managers. They allocate resources, including budget and manpower, to support energy efficiency initiatives. Superintendents and marine managers also collaborate with ship-owners, charterers, classification societies, and other stakeholders to promote energy efficiency and sustainability in maritime operations. Company superintendents and marine managers have an important role in implementing engine monitoring systems to enhance the energy efficiency of ships. Through their efforts, they drive improvements in fuel efficiency, reduce operational costs, and contribute to environmental sustainability in the maritime industry.

- **Technical Experts:** The data churned out by the EMS is often beyond the scope of ship officers and company personnel to make a complete analysis. Thus, shipping companies employ data analysts along with technical experts who study the data and confirm whether any improvements are needed. They include naval architects, marine engineers, mechanical and electrical engineers, energy management specialists, research scientists, and so on, who have a close understanding of the shipping industry and the ways and means for improving energy efficiency. Their expertise in system design, data analysis, performance optimisation, and maintenance, they contribute to significant fuel savings and environmental sustainability in the maritime industry.

- **Classification Societies and Recognised Organisations:** They conduct audits and surveys on board to ensure that the vessel is in compliance with mandatory and statutory requirements. In case of any non-compliance or shortfalls, they bring it to the knowledge of the ship's officers and the operators and suggest ways and means to correct these shortcomings. They are involved with the ship from "shipyard to scrapyard" and ensure the ship's structural integrity, seaworthiness, and overall readiness to withstand the challenges of the sea, while also safeguarding the environment. Classification societies and ROs play a vital role in facilitating the adoption and implementation of engine monitoring systems to improve energy efficiency and sustainability in the maritime sector. Through their standards, certification processes, technical support, and advice, they contribute to significant reductions in fuel consumption, emissions, and operational costs for ships worldwide.

- **Ship-Owners and Shipping Companies:** They play the most important role in ensuring the success of the EMS, resulting in increased energy efficiency. They play a proactive role in leveraging engine monitoring systems to enhance the energy efficiency of their vessels. Through investment, data analysis, training, and collaboration, they drive continuous improvement in fuel efficiency and environmental performance, ultimately contributing to a more sustainable maritime transportation industry.

The EMS is a vast system and it requires the hard work of both the ship and shore staff to ensure that the vessel is in compliance with regulatory requirements regarding energy efficiency. To assist them, there are outside agencies who have domain knowledge and technical know-how. Thus, operating a ship is a coordinated effort, and all parties must have adequate knowledge of the IMO regulations regarding fuel consumption and GHG emissions. There are several training modules available for this and both ship's officers and shore personnel are often required to attend these courses to get a holistic idea of energy efficiency.

5.4 IMPORTANCE OF MAINTENANCE: AN O-RING CAN INVITE TROUBLE

On one of the vessels where I was the master, the flag state control officers boarding our ship in an Indian port in 2018 observed black smoke emissions from the vessel behind our ship. Since our visit was pre-planned, they met me and requested to delay the flag state inspection since they had decided to visit the ship emitting the smoke. They returned to our ship after a couple of hours. During our inspection, they disclosed that the ship's engineers had no idea of the smoke emanating from their funnel. On being informed, they checked the diesel generator and found that the O-ring attached to the air cylinder had hardened and deteriorated. The return valve on the fuel shut-off was driven by the air cylinder. Due to the leakage, the valve could not close properly. Due to this, the back pressure in the line flowed back to the combustion chamber, which caused the air and fuel mixture to be unbalanced. This resulted in incomplete combustion and subsequent black smoke. Due to this, the vessel had to stop operations for more than six hours resulting in a financial loss to the ship operators. Furthermore, due to incomplete combustion, there was excessive fuel combustion and GHG emissions, which led to pollution of the environment.

The moral of the story is that routine maintenance is necessary for the health of the engines and generators. If not, such instances can occur, resulting in pollution of the environment, delay to the ship, and financial loss to the company. In your opinion, could this situation have been avoided?

5.5 SUMMARY

In this chapter, we have discussed the various aspects of monitoring and controlling fuel consumption. The methods of measuring fuel consumption are detailed, as well as the ways and means to reduce fuel consumption by looking after machinery health, performance monitoring, and so on. In addition, the details of the engine monitoring system (EMS) and highlighted. We have also looked at the role of the various persons involved in the ship operations related to the EMS.

Practically it is possible to substantially reduce fuel consumption by adhering to the basic principles of marine engineering – good maintenance. Systems have been set up for the same, such as PMS, it needs commitment by both the shore staff (supplying required resources) and the ship staff (regular inspections and maintenance) to ensure that the ship reduces her GHG emissions.

It is important to understand that the main reason for GHG emissions from ships is fuel consumption. Thus, by monitoring and reducing fuel consumption, we can reduce the emissions, giving us the benefits of both controlling global warming and complying with the mandatory requirements.

Having discussed the importance of fuel consumptions and monitoring the same we can move to the next chapter on low-carbon ship operations. We will see the carbon footprint of the ship and ways and means to reduce the same.

BIBLIOGRAPHY

18. IMO. (2019). *Data Collection System (DCS)*. https://www.imo.org/en/our work/environment/pages/data-collection-system.aspx.
    This explains the IMO data collection system, its timeline as well as its implementation.
19. Guidelines/Best Practices on Monitoring and Reporting Fuel Consumption by the European Sustainable Shipping Forum (ISSF). (2017). https://climate.ec.europa. eu/system/files/2017-07/02_guidance_monitoring_reporting_parameters_ en.pdf.
    This document published by the European Union details the guidance and best practices on monitoring, reporting, and verification of emissions from maritime transport.

Chapter 6

Low Carbon Ship Operations

The effort of global shipping is to reduce the emission of greenhouse gases (GHGs) into the atmosphere. Carbon dioxide is the principle GHG from ship's exhaust gases with results in global warming. The aim is to limit global warming to 1.5°C above preindustrial levels. Low carbon operations form the backbone of any initiative to reduce the impact of noxious emissions into the environment.

The International Maritime Organization (IMO) has issued several directives towards achieving sustainable development, with one of the most important being the reduction of sulphur in marine fuels to 0.5% m/m with effect from 1 January 2020, and 0.1% m/m within ECAs. This is one of the most significant steps towards reducing the emission of sulphur dioxide into the atmosphere. Further, the limits on emission of nitrous oxides into the atmosphere have also come into effect, aligning with the broader goal of environmental preservation and sustainable practices in the maritime industry.

The emission of carbon dioxide did not have any regulatory limitations so far. But with the new regulations of EEDI and Ship Energy Efficiency Management Plan (SEEMP) kicking in from 2013, this aspect will also receive the attention it deserves.

In this chapter, we shall discuss the importance of carbon intensity and the CII ratings. The low carbon operations for ships as well as the types of fuels to achieve the same are also discussed. This will give the reader a fair idea of how the IMO's final GHG study will be achieved.

6.1 THE IMPORTANCE OF CARBON INTENSITY

As explained in Chapter 3, carbon intensity is a major parameter to measure the carbon footprint of ships. Fossil fuels are carbon-intensive and their combustion releases carbon dioxide into the atmosphere, which ultimately results in global warming with alarming consequences.

Figure 6.1 shows the increase in atmospheric carbon dioxide over the years. As we can see, there is an exponential rise in CO_2 year on year. It is

DOI: 10.1201/9781032702568-6

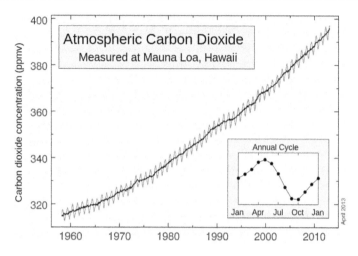

Figure 6.1 Increase in atmospheric carbon dioxide.

Source: Wikimedia (2013)[20]

hoped that with the new regulations and stringent IMO strategy, CO_2 emissions will gradually decrease.

According to Climate.gov,[21] the atmospheric carbon dioxide was 417.06 parts per million (ppm) in 2022. For the past 11 years, this has been increasing at a rate of more than 2 ppm annually.

The CII, which we introduced in Chapter 3, is considered to be a measure of the energy efficiency of the ship, that is, the reduction in fuel consumption; it is an indicator of the CO_2 emission of the ship and is directly proportional to the fuel consumed. It is given by the grams of CO_2 per transport mile, where the transport mile is the ship's cargo carrying capacity (DWT) × design speed. As per MARPOL Annex VI amendments, ships have to determine their annual CII.

As per amendments to MARPOL Annex VI, carbon intensity measures have come into effect from 1 January 2023, as follows:

- The CII is to be calculated for the year 2023.
- The operators are required to reduce the CII year on year.
- The operational CII will be calculated for 2024.
- The vessels will receive an environmental rating as follows:
  - A for a major superior performance in reducing the CII.
  - B for a minor superior performance.
  - C for a moderate performance in reducing the CII.
  - D for a minor inferior performance.
  - E for an inferior performance level.

These ratings will be recorded in the "Statement of Compliance". The first annual reporting will be completed by the end of 2023, and on the basis of this, the CII ratings will be given in 2024.

The 2023 IMO GHG strategy requires a reduction in carbon intensity by at least 40% by 2030. In order to achieve this, new technologies, alternate fuels, and so on should account for at least 5% of the consumption of fuel in international shipping while striving for 10%. Only then will the IMO ambitions be achieved.

IMO Resolution MEPC 377 (80), adopted on 7 July 2023, outlines the 2023 IMO GHG strategy on GHG reduction.[22] This resolution contains the levels of ambition related to the 2023 IMO GHG strategy dealing with the carbon intensity of the ship and the indicative checkpoints to reach net-zero emissions. MEPC 80 is a landmark resolution paving the way for achieving zero-carbon emissions from ships in 2050. It will be discussed in detail in Chapter 9.

IMO has been concerned with the fuel consumption on board ships, which results in GHG emissions causing global warming. As we start reducing fuel consumption on board due to the IMO strategy, it is necessary to look at the emissions in totality. This concept of measuring GHG emissions from the very beginning to the very end is known as "well to wake", that is, from the oil well where the fuel is extracted to the wake of the moving ship.

6.1.1 Well to Wake

This is the term used to describe the entire process of fuel consumption and GHG emissions starting from fuel production to the ship's exhaust gases. It is known as the life cycle assessment (LCA). Conducting an LCA for ship fuels entails analysing the environmental consequences of a fuel across its complete life cycle. This encompasses every stage, starting from the extraction of raw materials and the production of the fuel to its transportation, utilisation in ships, and eventual disposal or recycling. The objective is to offer a thorough comprehension of the environmental impact associated with various categories of ship fuels.

Well to wake has two parts:

- **Well to Tank:** This includes the extraction, processing, and refining, as well as the transport and distribution of the fuel.
- **Tank to Wake:** This refers to the combustion on board resulting in the emission of exhaust gases.

IMO has issued guidelines on the life cycle GHG intensity of marine fuels[23], also known as LCA guidelines. Life cycle assessment means the assessment of GHG emissions from well to wake. It includes the emissions from the extraction process, the processing and refining of the fuel, the transportation to the final destination, and finally, the combustion of the fuel on board the ship.

So far, we have seen how carbon dioxide and GHG emissions from ships lead to global warming. Low-carbon ship operations can help address this issue. Let's look at what these operations are and how they can be achieved.

6.2 SO WHAT ARE LOW-CARBON SHIP OPERATIONS?

This refers to the need to reduce the carbon footprint of the vessel in order to limit global warming and meet the sustainable development goals of the United Nations. Low-carbon ship operations are possible by adopting practices, technologies, and strategies aimed at reducing the carbon footprint and GHG emissions from ships. These initiatives focus on minimising the environmental impact of shipping activities while maintaining or improving operational efficiency and safety. They include various aspects of vessel design, propulsion systems, fuel choices, operational practices, and regulatory compliance.

To enhance the low carbon operations on board, the following aspects are to be considered:

- Alternate fuels are one of the most important aspects to reduce the carbon footprint of ships. Alternate fuels in consideration are those that do not emit CO_2. Fuels like LNG emit less CO_2 than fossil fuels like heavy fuel oil and marine diesel oil but still emit GHGs. Fuels like green methanol, green hydrogen, and so on are the real zero-carbon options for ships. Alternate fuels will be discussed in detail later in this chapter.
- Use of renewable energy is another zero-carbon option for ships. Development and research for renewable energy sources are in progress. Wind energy, solar energy, and even tidal energy are being actively considered as fuel sources for ships.
- Hull maintenance is necessary to reduce frictional losses and improve fuel efficiency. Superior hull coatings, underwater hull cleaning, propeller polishing, and so on are some of the measures to achieve low-carbon operations.
- Operational practices to reduce fuel consumption and emission of GHG. This has already been discussed in earlier chapters.

To ensure low carbon operations, it is important to use the correct type of fuel. Let's understand why.

6.3 TYPES OF FUEL FOR REDUCTION OF GHGs

The type of fuel used in ships is a critical factor in reducing emissions from the ship's exhaust. Let us have a look at the various fuels that can be used to power ships and consider their advantages and disadvantages.

6.3.1 Heavy Fuel Oil

HFO is the fuel used on most ships. Due to the cost-effectiveness, it is difficult to replace HFO as the fuel for ship's engines till equivalent alternatives are found. HFO contains sulphur and emits CO_2 during combustion. The emission of SOx has been curtailed by the stringent requirements of IMO, which capped the sulphur content in marine fuels at 0.5% m/m outside ECA and 0.1% inside ECA.

HFO is one of the most carbon-intensive fuels, but reduction of carbon dioxide is not feasible. The solution is to use alternate fuels and sources of energy in order to minimise CO_2 emissions. Currently, the use of alternate fuels is at an early stage, and it is hoped that the production of these alternate fuels will pick up, lowering the cost and making them more attractive for ship-owners. Due to economic and commercial reasons, it is not possible to enforce a changeover to these fuels until viable alternatives are available in the market.

Currently, HFO must not exceed a sulphur content of 0.5% m/m (mass by mass) outside ECA and 0.1% m/m within ECA (which is known as "ultra-low sulphur"). Violation of these limits may lead to severe penalties and fines by port state control and other port agencies.

The chief engineer must keep a record of the changeover to ultra-low sulphur fuel before entering any designated ECA and maintain a record of the same. The changeover procedure must be started 12 hours before entering such areas.

Port state control officers within ECAs often board ships to check for compliance with changeover procedures to ultra-low sulphur fuels. Let's look at an example. In one such case, the chief engineer had forgotten to change over to ultra-low sulphur fuel before berthing in a European port. The port state control officer on boarding the vessel identified this non-compliance. A sample of the fuel oil was collected from the pipeline and sent to the lab for analysis. The fuel oil was found to have a sulphur content exceeding the permissible limits. A deficiency was issued to the vessel, and the ship-owner had to pay a hefty fine for the vessel to be released. The vessel meanwhile had to use shore power for ship operations, and a huge cost was incurred for this service. This incident highlighted the need to follow proper procedures and maintain evidence of compliance in order to avoid unnecessary deficiencies, penalties, costs, and detentions by port state control.

6.3.2 Marine Diesel Oil

This is a relatively cleaner option compared to HFO. It contains less sulphur and produces lower carbon emissions per unit of energy. MDO is used on marine engines using HFO during manoeuvring, berthing, and so on. MDO is a lighter and more refined fuel as compared to HFO and has a lower viscosity and fewer impurities. It emits less GHG including CO_2 and SOx than HFO. The carbon footprint of MDO is lesser than that of HFO. However, due to its higher cost, vessels prefer to use HFO during long passages. There are also no regulations for the use of specific fuels as long as the SOx emissions are as per regulations.

Currently, the majority of ships operate on HFO for ocean passages and MDO for manoeuvring. The use of MDO for long passages is avoided because of the cost factor. Ultimately the choice between HFO and MDO is influenced by many interlinking factors such as fuel prices, emission control, ship's engine suitability, and so on.

6.3.3 Alternate Fuels

It has been long established that conventional fuels such as HFO and MDO contribute to the carbon footprint of the ship. Thus, the search for alternate fuels with a lesser carbon footprint has been going on for a long time. Although there are a number of alternatives, their effectiveness as well as the costs involved are proving to be prohibitive for the ship-owner. The choice of fuel depends on factors such as vessel type, operational profile, regulatory compliance, and the availability of infrastructure and support mechanisms.

Let us have a look at the alternate fuels and their usefulness:

6.3.4 Liquefied Natural Gas

LNG produces lesser emissions of CO_2, SO_x, particulate matter, and so on, as compared to HFO and MDO. It is less polluting, and this makes it a better option for use as fuel to meet regulatory compliance. It has a high energy density, and its use as a ship's fuel will increase energy efficiency, resulting in lower fuel consumption and reduced costs. It can be used in dual-fuel engines and is thus an attractive transition fuel to lower the environmental impact of the ship. Additionally, they produce less noise and vibration as compared to conventional engines.

LNG is also a fossil fuel but reduces CO_2 emissions by about 20%.

The main disadvantage of using LNG is methane slip, the escape of unused methane from the engine's combustion chamber. Methane slip also occurs in areas where it is stowed. Ultimately, LNG being a fossil fuel, these are concerns about the long-term sustainability.

Retrofitting existing ships to run on LNG or building new LNG-powered vessels requires significant modifications to the propulsion systems and on-board infrastructure, which can prove to be expensive for the ship-owner. LNG possesses a reduced energy density relative to conventional marine fuels such as diesel or heavy fuel oil. Consequently, vessels fuelled by LNG may necessitate either enlarged fuel storage capacities or reduced operational ranges between refuelling. LNG can be considered as a short-term measure to reduce GHG emissions till long-term solutions are found. Further, LNG engines are more efficient and can achieve energy savings resulting in reducing the carbon footprint of the ship.

Bio-LNG is the latest alternative for carbon-intensive fuels, and research is in progress to determine its effectiveness on board and the scope for production. Bio-LNG can be produced from organic waste materials by biomethane liquefaction. Biogas is generated from the organic waste, purified, and then liquefied to convert it to bio-LNG. Since it does not involve any fossil fuel, it has a much lesser carbon footprint than conventional LNG.

Since conventional LNG is not carbon neutral, it will be phased out as other alternatives pick up. This is the reason LNG is a transition fuel towards carbon-zero shipping.

6.3.5 Biofuels

Biofuels have a lower carbon footprint as compared to fossil fuels. It is therefore considered an attractive option to comply with IMO's GHG strategy. It is expected that biofuels will power at least 30% of the global fleet over the next few years. There are several types of biofuels that are considered suitable for ship's engines:

- **Fatty Acid Methyl Ester (FAME)** is produced from waste cooking oils, vegetable oils, or animal fats via a process known as transesterification (a chemical conversion process). It is also known as biodiesel.
- **Hydrogen vegetable oil/Hydrogenation-derived renewable diesel (HVO/HDRD)** is the product of vegetable oils or fats either on their own or blended with petroleum fuels. These are also known as renewable diesel or green diesel. They are a fossil-free option resulting in reduction of GHGs by almost 90%.
- **Straight Vegetable Oils (SVO)** are produced from rapeseeds, soybeans, or palms. It is considered superior to FAME and HVO since it is a renewable fuel derived directly from plants. Thus it is less processed and has a lower CII.
- **Biomass to Liquid (BTL)** is produced from biomass by thermoschemical conversion. These are synthetic fuels that are chemically different from conventional fuels but can be used in diesel engines with some modifications.

All the above biofuels have their own advantages and disadvantages. But they share a common advantage, that of reducing the carbon footprint of the ship. It is up to the ship-owner or operator to choose the type of fuel they wish to use.

The CO_2 released during the combustion of biofuels is part of the natural carbon cycle. The plants used to produce biofuels absorb CO_2 from the atmosphere during their growing stage and thus offset the CO_2 released during combustion. Although biofuels do release CO_2, they can be considered carbon neutral because they have already absorbed CO_2 during their growth stage.

But the use of biofuels for maritime transportation comes with its own set of disadvantages. Since many biofuels are derived from agricultural crops, their use as fuel may divert resources away from food production, leading to increased food prices.

To combat this, advanced biofuels are being produced using technological processes. Second-generation biofuels can be produced from agricultural residues, algae, woody biomass, and so on. These offer even more carbon reduction than first-generation biofuels. In addition, demand for these agricultural products provides economic upliftment and supports rural economies, contributing to sustainable development.

ClassNK has published concise details about biofuels. It details the type of biofuels in use, the statutory requirements, and precautions to be taken when using biofuels. "Handling and use of biofuels on ships"[24] (Class NK, n.d.) has an FAQ that attempts to clarify any doubts about biofuels.

6.3.6 Methanol

Using methanol as fuel on ships is an option for the ship-owners and operators to reduce the carbon footprint of their ships. Methanol (CH_3OH or MeOH) is a colourless, biodegradable alcohol used extensively in the paint and pharmaceutical industry. It is toxic and flammable but dissolves in water. Combustion using methanol emits less carbon dioxide than fossil fuels and hence the industry of the shipping world on this fuel as an option to reduce the ship's carbon footprint in order to comply with regulatory requirements of EEDI and EEXI.

The amount of reduction of CO_2 emissions depends on the type of methanol used.

Methanol can be categorised into fossil-based, produced from coal or natural gas, and renewable methanol from biomass or captured CO_2 with green hydrogen. The following are the types of methanol:

- Brown methanol which is produced using coal. The coal is gasified to produce synthesis gas (syngas), a mixture of carbon monoxide and hydrogen. The syngas is then synthesised to produce methanol. It does not contribute much to the reduction of CO_2 emissions, as CO_2 is released during the production process, and it is not considered an alternate fuel for ships.
- Grey methanol, which is produced using natural gas which is primarily methane (CH4). It is subjected to a procedure called steam methane reforming (SMR) to produce hydrogen and carbon monoxide. The carbon monoxide is then synthesised with additional hydrogen to form methanol. Like brown methanol, it does not contribute to much reduction of GHG and is not considered as ship fuel.
- Green methanol can be said to be the zero-carbon fuel as it does not involve any emissions of GHG during the manufacturing process as

well as during combustion on board the ship. There are two types of green methanol depending on their origin.

Bio methanol is one of the most sustainable options since it is produced from renewable sources, as it is produced from biomass. Feedstock such as sugarcane residue is used to produce bio methane. The biomass is gasified to convert it into a mixture of hydrogen, carbon monoxide, and other products. The syngas is then purified and synthesised to produce methanol (CH_3OH). But the challenge is the availability of large quantities of feedstock for bio methanol production. However, methanol production is on the upswing and is expected to cater to a substantial percentage of the world's fleet by 2050.

- E methanol is produced through a combination of green hydrogen, captured CO_2, and renewable electricity. Hydrogen is produced in an electrolyser. The hydrogen is then converted to methanol by catalysis with CO_2 in a reactor. It is considered a sustainable option because it is carbon neutral.
- Green methanol is the umbrella term used for bio methanol and E-methanol. They mean the same, that is, methanol which has zero carbon.
- Blue methanol is produced from blue hydrogen and captured CO_2. Use of this as ship's fuel reduces CO_2 emissions, but it is not carbon neutral.

Methanol has a major disadvantage in that it can cause serious health effects if inhaled or ingested, including respiratory irritation, headaches, nausea, dizziness, and in severe cases, blindness or even death. Thus, all safety protocols must be in place to ensure the safety of the crew on board the ships.

Overall, the use of methanol as fuel can reduce the carbon footprint of ships to a large extent. As availability increases, more ship-owners would be willing to take an extra step to reduce the carbon emissions from their ships.

For example, A.P. Moller-Maersk has taken delivery of the 32,300 dwt, 2100 TEU, 172m long feeder vessel M.V. Laura Maersk in July 2023. It is a dual-fuel engine that can sail on green methanol. There are 19 such vessels on order which can save around 2.3 million tonnes of CO_2 emissions annually.

The Methanol Institute[25] is the global association for methanol producers and distributors. One of the disadvantages of using methanol in the future could be global availability as demand increases. To alleviate this, the Methanol Institute aims to streamline supply and improve supply chain management.

6.3.7 Ammonia

Ammonia (NH3) can be used as an alternate fuel to the carbon-intensive HFO and MDO. Ammonia has safety risks due to its toxicity and corrosivity to humans. Ammonia has a low energy density and thus large amounts of it have to be carried on board as compared to conventional fossil fuels.

There are several types of ammonia depending on the source of their production.

- Brown ammonia is produced by the gasification of coal to produce synthesis gas (syngas) containing hydrogen and carbon monoxide. This is then combined with nitrogen to produce ammonia, which is then purified to make it ready for use. Since coal is the source of ammonia, CO_2 and other gases are released during the production of brown ammonia, and hence it is not considered as an alternate fuel for ships.
- Grey ammonia is produced from natural gas by a process called natural gas reforming. During this process, CO_2 is released into the atmosphere and is a significant contributor to GHG emissions. Hence, it is not suitable as an alternative to HFO or MDO.
- Blue ammonia is also produced from natural gas by a process called SMR, which produces hydrogen gas and carbon dioxide. CO_2 is then made to react with steam and produce more hydrogen and carbon monoxide (CO). The hydrogen is then synthesised to form ammonia, while the CO_2 is captured and stored. While blue ammonia is a lower-carbon alternative as compared to fossil fuels, it is not a zero-emission solution, as some amount of carbon is released due to inefficiencies in the production and the energy requirements for carbon capture and storage.
- Green ammonia is a zero-carbon fuel and thus will have a key role to play in meeting the challenges of the GHG strategy of IMO. It is synthetically produced by electrolysis. It has all the desired advantages but is expensive compared to other fuels. It is hoped that with increasing production, the cost will come down and green ammonia will become more affordable. It is produced by performing electrolysis on water which then breaks into hydrogen and oxygen. The hydrogen is then made to react with nitrogen from the atmosphere to produce ammonia.

There is no doubt that green ammonia is the fuel of the future. As per MAN Energy Solutions, ammonia is expected to account for approximately 27% of fuel mix for large ships.

MAN B&W, a renowned marine engine manufacturer, has announced the successful test of their two-stroke engine on ammonia.[26] Their efforts at improving the safety and fuel efficiency of these engines have been successful.

6.3.8 Hydrogen

The use of hydrogen as a fuel for the ship is an attractive option as hydrogen is an alternate fuel. There are mainly four types of hydrogen that can be produced:

- Brown hydrogen is produced from coal-by-coal gasification. Coal reacts with oxygen and steam to produce synthesis gas (syngas).

The syngas is then made to undergo a water-shift reaction by adding water vapour to produce more hydrogen and carbon dioxide. The carbon dioxide is then either released to the atmosphere or captured by carbon capture and storage (CCS). But even with CCS this process releases CO_2 and other pollutants into the atmosphere and so is not a sustainable option as a marine fuel.

- Grey hydrogen is produced by extracting methane from natural gas and subjecting it to SMR. Hydrogen gas is produced when the methane reacts with steam in the presence of a catalyst. Grey hydrogen is not a zero-carbon option, as carbon dioxide is one of the by-products of the production process of grey hydrogen, mainly during the SMR process. Thus, grey hydrogen is not suitable as an alternate fuel for ship's engines.

- Blue hydrogen is also extracted from natural gas in a similar way to the production of grey hydrogen. However, in the production process of blue hydrogen, the carbon dioxide released during the SMR is captured and transported for storage or utilisation in other industrial processes. Blue hydrogen with carbon capture reduces CO_2 emissions to a large extent, but the entire process does involve CO_2 emission into the atmosphere. Blue hydrogen can, at best, be used as a transitional fuel while the technology for green hydrogen production is being streamlined.

- Green hydrogen is a zero-carbon fuel and can be termed as the fuel of the future. It is produced by the electrolysis of water into hydrogen and oxygen. The electrolysis is powered by renewable energy such as wind, solar, geothermal energy, and so on. Technology is being developed to make this process more efficient and cost-effective.

Of these, green hydrogen is the best option as it is a zero-carbon fuel. At present, it is in the development stage and is being used for smaller craft such as ferries, small boats, and so on. But shipyards have received orders and are in the process of building larger ships powered by hydrogen cells.

The major disadvantage of hydrogen as a ship fuel is the safety concerns due to its flammability and high reactivity. Hydrogen has a wide flammability range and can ignite at concentrations as low as 4% in air, making it more prone to accidental leaks or explosions if not properly managed. Thus, safe storage, bunkering, and handling procedures are necessary for hydrogen to be widely accepted as a fuel for ships.

The American Bureau of Shipping (ABS) has released a sustainability whitepaper titled "Hydrogen as Marine Fuel".[27] It enumerates the advantage of using hydrogen as a marine fuel in order to reduce the GHG emissions from ships. The paper explains the safety hazards associated with the use of hydrogen as a marine fuel and the ways to combat this.

So what is the future of hydrogen? It is a clean energy source, abundantly available all around us. Apart from the safety factor, which the shipping industry is working around, another disadvantage is the cost factor, but it is hoped that with new technology, increasing demand, and the resulting economies of scale, hydrogen will become cheaper.

The International Energy Agency (IEA) has published a report titled "The Future of Hydrogen[28]" (June 2019), which deals with the demand for hydrogen as marine fuel, the industry support for hydrogen as fuel, as well as the cost factor in its production.

In order to study the effects of GHGs so that they can bring in updated legislation, IMO has been conducting GHG studies since 2020. The aim is to estimate GHG emissions and project future emissions. The latest in this series of GHG studies is the fourth GHG study.

6.4 IMO FOURTH GREENHOUSE GAS STUDY

IMO conducted their fourth GHG study[29] in 2020 and published the key findings as follows:

- The total GHG emissions of total shipping have increased from 977 million tonnes in 2012 to 1,076 million tonnes in 2018, that is, an increase of 9.6%.
- The share of shipping in global emissions has increased from 2.76% in 2012 to 2.89% in 2018.
- Specifically, CO_2 emissions have increased from 701 million tonnes in 2012 to 740 million tonnes in 2018, that is, 5.6% increase.

The fourth GHG study has exposed the increase in carbon emissions over time. If action is not taken urgently, then it would not be possible to limit global warming to 1.5°C above preindustrial levels as required by IMO. This is the reason all stakeholders must take all possible measures to limit their CO_2 emissions and reduce the carbon footprint of the ships.

6.5 SUMMARY

In this chapter, we have seen that ships have to record their fuel consumption and report the same to the administration or RO. Based on their fuel consumption, ships will be rated as A, B, C, D, or E. In order to maintain the CII ratings as A or B, vessels have to lower their dependence on fossil fuels and thereby reduce their CO_2 emissions. Low-carbon ship operations are the backbone of IMO's strategy to reduce GHG emissions and thereby limit the global warming to 1.5% above preindustrial limits. Ship-owners and other stakeholders have to commit to increasing the fuel efficiency of the ships by reducing fuel consumption and associated GHG emissions. The various alternate fuels are discussed along with their advantages and suitability for ships. By having a clear grasp of the requirements of low-carbon operations of ships, you will understand why energy efficiency is important and how this can be achieved.

In the next chapter, we shall see how the efficiency of marine diesel engines can be increased to reduce fuel consumption. It also discusses the various renewable energy sources available as alternatives to diesel engines.

BIBLIOGRAPHY

20. Increase in Atmospheric Carbon Dioxide. (2013). https://commons.wikimedia. org/wiki/File:Mauna_Loa_Carbon_Dioxide_Apr2013.svg.
    This graphic shows the increasing levels of atmospheric carbon dioxide over the years.
21. Understanding Climate Change. (2023). https://www.climate.gov/news-features/ understanding-climate/climate-change-atmospheric-carbon-dioxide.
    This article by Rebecca Lindsay, in Climate.gov, takes a look at the increase of atmospheric carbon dioxide over the years. It becomes abundantly clear that if urgent action is not taken, the presence of carbon dioxide in the atmosphere will increase exponentially resulting in catastrophic climate change.
22. Resolution MEPC 388(80). *2023 IMO GHG Strategy on Reduction of GHG Emission*. https://wwwcdn.imo.org/localresources/en/MediaCentre/PressBrief ings/Documents/Clean%20version%20of%20Annex%201.pdf.
    This resolution commonly known as MEPC 80 mandates the IMO requirements for reduction of GHG emissions from ships. It states the levels of ambition, short-term, mid-term and long-term measures for GHG reduction with timelines.
23. Guidelines on Life Cycle GHG Intensity of Marine Fuels (LCA Guidelines). (2019). https://www.imo.org/en/OurWork/Environment/Pages/Lifecycle-GHG-carbon-intensity-guidelines.aspx.
    This report explains the life cycle assessment methodology as well as the development of guidelines on life cycle GHG intensity of marine fuels.
24. Understanding the Use of Biofuels on Ships (ClassNK). (2023). https://www. classnk.or.jp/hp/en/info_service/bio/.
    This document by Class NK looks into the use of biofuels and contains an FAQ for better understanding of the use of biofuels on ships.
25. Methanol Institute. (2024). https://www.methanol.org/about-us/.
    Methanol Institute is the global association for methanol producers, distributors and technology companies.
26. Ammonia Engines by MAN Energy. (2023). https://www.man-es.com/company/ press-releases/press-details/2023/07/13/groundbreaking-first-ammonia-engine-test-completed.
    In this document MAN B&W announced the successful test of the first ammonia engine.
27. ABS Report on Hydrogen as Marine Fuel. (June 2021). https://ww2.eagle.org/ content/dam/eagle/publications/whitepapers/hydrogen-as-marine-fuel-white paper-21111.pdf.
    This whitepaper essentially addresses the safety considerations of using hydrogen as a marine fuel.

28. International Energy Agency Report on Future of Hydrogen. (June 2019). https://www.iea.org/reports/the-future-of-hydrogen.

This report by the IEA discusses the future of hydrogen as a fuel in various industries globally. It also states the seven recommendations of IEA to scale up hydrogen production, so as to make it easily available.

29. IMO Fourth Greenhouse Gas Study in 2020. https://wwwcdn.imo.org/localresources/en/OurWork/Environment/Documents/Fourth%20IMO%20GHG%20Study%202020%20-%20Full%20report%20and%20annexes.pdf.

This detailed document by IMO published in 2020 covers the fourth GHG study conducted by IMO as a result of the initial IMO GHG strategy. It covers the status of emissions as well as the carbon intensity globally. It also gives the projection and estimates of future emissions.

Chapter 7

Diesel Engines and Energy Efficiency

The last two decades have witnessed an exponential increase in the size and speed of ships. Container vessels have been in the forefront of this race for larger ships. From 5,000 TEU (tons equivalent units) to 10,000 and now 22,000 TEU, the rise has been astronomical. While economies of scale have made this transition profitable for the ship-owner, the environment has been at the receiving end. Because of their size and speed, Post-Panamax vessels consume much more fuel than bulk carriers and tankers.

In this chapter, we will discuss the pros and cons of marine diesel engines using fossil fuels. We will also see the technological developments for reducing the carbon footprint of ships, as well as the retrofitting of advanced technology on existing ships. Renewable energy is the need of the hour, and we shall have a look at the optimum solution to greenhouse gas (GHG) emissions by use of renewable energy such as solar energy and wind energy.

Most of the ships sailing the seas today are fitted with diesel engines which use heavy fuel oil (HFO) or marine diesel oil (MDO) as their source of energy. There is no doubt that fossil fuels are indeed the most popular choice for ship engines. More than 90% of ships operate on fossil fuels. Let us have a look at why this is so.

7.1 WHY ARE FOSSIL FUELS POPULAR FOR SHIPS' ENGINES?

Fossil fuels like HFO and MDO are popular among ship-owners because of their many advantages:

- **Cost Benefit:** HFO and MDO are no doubt the most cost-effective fuels today. In addition, diesel engines are more fuel-efficient than other engines, resulting in lower fuel consumption.
- **Range:** As they are fuel efficient, diesel engines are suitable for ships where the voyages across oceans last many days. Further, the easy availability of HFO and MDO across the world makes them an attractive option.

DOI: 10.1201/9781032702568-7

- **Maintenance:** Diesel engines are easier to maintain than other engines and have a longer life span. They are generally of robust construction and simple design. This leads to longer service intervals and ease of on-board maintenance by the ship's engineers. Since diesel engines are popular and widely used, their parts are available at most ports, and therefore the replacement of damaged or worn-out parts becomes easier.

Marine diesel engines also come with their own disadvantages, which include the following:

- **Emissions:** This is one of the major disadvantages of diesel engines. Diesel engines emit greenhouse gases (GHGs) which are damaging to the global environment, as well as the health of humans, agriculture and oceans. Thus, International Maritime Organization (IMO) has brought into force several measures to limit such emissions. The IMO strategy is to finally phase out fossil fuels altogether and encourage the use of zero-carbon fuels.
- **Maintenance Cost:** As per the International Safety Management (ISM) Code, ships must implement the PMS on board and ensure this is followed. The PMS requires regular oil changes, filter replacements, regular surveys, and replacement of worn-out parts. All these add to the maintenance costs of the engines.
- **Noise and Vibration:** The process of producing power in diesel engines produces combustion noise, which is amplified due to resonance in the engine room. In addition, these engines produce mechanical noise and vibration due to the various moving parts. The vibrations are transmitted to the hull and felt all over the ship. Though the noise and vibration cannot be removed altogether, they can be reduced to some extent by proper maintenance.

Despite the disadvantages, diesel engines will continue to be in use for a long time. Fossil fuels remain the preferred fuel for ships due to their high energy density, cost-effectiveness, global availability, and easy transportability, which make them convenient for shipping operations worldwide. With significant energy per unit of weight, they enable long-distance travel without frequent refuelling. Although there is an increasing interest in alternate fuels mainly due to regulatory pressures, fossil fuels remain the fuel of choice for most ship-owners. But new ship orders are seeing an increasing trend away from fossil fuels, which will lead to a more sustainable future.

Then what is the solution for a ship operator who wants to continue using fossil fuels due to its many advantages? The solution is to bring down the exhaust emission within the permissible range. At higher speeds, the fuel efficiency is lower, the fuel consumption is higher and thereby the emission of GHG and other harmful gases is higher. Operating the engines within the optimum operating range will reduce emissions and increase fuel efficiency.

We have already seen how the conventional diesel engines operating on HFO/MDO can be made environment-friendly. In our chapter on best practices, we shall discuss the ways and means to reduce fuel consumption and thereby reduce GHG emissions. But the fact remains that despite these measures, ships operating on HFO/MDO are unlikely to meet the IMO strategy of GHG reduction. Thus, there is a crying need to find alternatives and shift away from the use of fossil fuels. Unfortunately, although technology has invaded every part of our lives, it is a late arrival as far as ships and specifically ship engines are concerned. However, given the current thrust on the reduction of GHG emissions, ship owners and engine manufacturers are looking to technology for solutions.

7.2 TECHNOLOGICAL DEVELOPMENTS IN MARINE DIESEL ENGINES

Ships have a life of between 25 and 30 years. Ships ordered during the last decade may well go past the IMO ambition of zero carbon by 2050. These vessels, operating on conventional fossil fuels, will continue to emit GHGs. The need of the hour is to integrate technological developments in these diesel engines so that their emissions are reduced to the minimum.

As per the latest version of the Maritime Forecast to 2050[30] published by Classification Society DNV, 51.3% of the current shipping order book is capable of running on alternate fuels and battery power. Of this, 40.3% can use LNG, 8.01% methanol, 2.24% LPG, and 0.8% battery power. Currently, only 6.52% of the global fleet can run on these fuels.

Marine engine manufacturers are constantly trying out new innovations to make their engines in compliance with stringent emission regulations. Technological developments are both feasible and cost-effective for new ships. With the passage of time, new ships operating on alternate fuels or renewable sources of energy will become popular. For existing ships, such advanced equipment needs to be retrofitted and is expensive as well as time-consuming to install. But there are several relatively advanced equipment which can also be fitted on existing ships. Let us have a look at some of these.

7.2.1 What Are the Options for Existing Ships?

Existing ships and ships not using alternate fuels or renewable sources of energy will have to look to incorporating technological developments to reduce their exhaust gas emissions. The following are the options available and being actively considered by the shipping industry:

- **Exhaust Gas Recirculation (EGR):** EGR systems recirculate a part of the exhaust gas back to the engine's combustion chambers to lower the combustion temperature and thereby reduce NO_x emissions. EGR systems are also helpful as the recirculation reduces the amount of exhaust gas emissions in general.

- **Selective Catalytic Reduction (SCR):** This technology is being used to reduce nitrogen oxide emissions. A catalytic converter along with a reducing agent converts the NO_x into nitrogen, water, and carbon dioxide. The common reducing agent is aqueous urea, which is injected into the exhaust gas stream. This passes through the catalytic converter, which breaks down the NO_x. The catalyst used is generally titanium oxide or vanadium pentoxide.

- **Organic Rankine Cycle (ORC) System:** This system is generally used on ships for WHR. The heat which would have been lost to the environment is recycled for additional power. An organic fluid with a lower boiling point than water is pumped to a low-pressure heat exchanger. In the heat exchanger, transfer of heat takes place between the exhaust gas and the organic fluid, which undergoes vaporisation. The organic fluid is generally hydrocarbons or refrigerants. A turbine converts the heat energy into mechanical work which is then converted to electrical energy by a generator. This electrical energy is then used to power the ship's auxiliary equipment or connect to the ship's electrical grid. The organic fluid is then condensed back to the fluid state and reused. This system recovers the waste heat and increases the fuel efficiency of the ship.

- **Turbo Compound Systems:** This is another type of WHR system where a turbine is integrated into the ship's exhaust system. The hot exhaust gases cause the turbine to rotate, which drives an electrical generator. The electrical energy so produced can be used to meet the ship's electrical load. The turbo compound system is much simpler than the ORC system but is less effective because of the temperature range stability and thermodynamic efficiency of the organic fluid.

By implementing a combination of these measures tailored to the specific characteristics and operational profile of the vessel, ship-owners and operators can achieve substantial improvements in energy efficiency. This will reduce fuel consumption, resulting in major cost benefits, reduce emissions resulting in sustainable operations and also will ensure regulatory compliance.

7.2.2 Engine Power Limitation and Shaft Power Limitation

Engine Power Limitation (EPL) and Shaft Power Limitation (ShaPoLi) are two innovative systems to ensure that the ship's engines are not operated at a higher-than-required revolutions per minute (RPM). The purpose is to limit the RPM and thereby fuel consumption. It has been scientifically proved that there is an exponential relationship between the speed of the ship and fuel consumption. Thus, power limitation is an acceptable form of ensuring that the speed of the ship is not exceeded to an extent where the fuel consumption increases drastically.

EPL is the optimum power that the ship's engines are designed to produce. This limitation is incorporated into the engines such that under normal operations they cannot operate beyond this. The limitation is set keeping in mind the safe speed of the ship in compliance with the regulatory requirements. Exceeding this limitation will exponentially increase fuel consumption and the EEXI requirements. However, in special cases, the master and the chief engineer can override the EPL. This can be done for weather considerations, maintaining the estimated time of arrival (ETA), navigational exigencies, or to escape from pirates.

ShaPoLi refers to the maximum power that can be transmitted through the propeller shaft to the propeller. This is essential not only for controlling fuel consumption but also to ensure that overloading and subsequent damage to the propulsion system is avoided. Exceeding the ShaPoLi can cause damage to the propulsion system, including the bearings, shafts, and so on.

The main difference between EPL and ShaPoLi is that while EPL relates to the maximum power output of the engines, ShaPoLi refers to the maximum power transmission to the propulsion system. Both achieve the same objective of limiting fuel consumption and ensuring that the propulsion system and the engines are not overloaded to an extent that can cause damage to the machinery. Thus, both energy efficiency and safety are taken care of by these limitations.

Let us now have a look at some advanced engine designs that are being considered for new ships. Engine makers are well aware of the need to reduce fuel consumption and are innovating new engine designs.

7.2.3 Advanced Engine Design for New Ships

For new ships in the process of construction, engine makers offer advanced engine design. This is another option for manufacturers to meet the tough emission standards. There are many solutions available to increase the energy efficiency of the engines:

- **Improving Combustion Efficiency** is one of the primary methods for increasing the efficiency of the engines. There are many methods that engine manufacturers adopt to improve combustion efficiency:
  - Optimising fuel injection systems to deliver the precise amount of fuel at the right time will improve fuel combustion. Upgrading injectors and adjusting the injection timing are some of the options available.
  - Improving the design of combustion chambers by improving its geometry will ensure uniform fuel distribution within the chamber. This is achieved by designing the chamber's dimensions and shape in order to facilitate optimal fuel-air mixing, resulting in better combustion.

- Better air-fuel mixture can also be achieved by creating finer fuel droplets by means of advancements in fuel atomisation technology and optimising airflow patterns within the chamber.
- **Advanced Engine Control Systems** are used in order to optimise engine performance. These systems employ advanced sensors, actuators, and control algorithms to monitor and adjust various parameters of the engines and propulsion systems. Engine speed, temperature, pressure, fuel flow, exhaust emissions, and shaft RPM are monitored, and algorithms are used to analyse this data in real-time to optimise the engine performance. Optimisation includes controlling fuel injection timing, adjusting air-fuel mixture ratios, adjusting turbocharger boost pressure, and managing exhaust gas recirculation systems.
- **Carbon Capture and Storage (CCS)** is another exciting innovation to reduce exhaust gas emissions from ships. This topic will be discussed in detail in later chapters.
- **Deactivation of Certain Cylinders** under light loads will allow for lesser fuel consumption. Some newer engines are designed to allow this, and these engines have better fuel efficiency.
- **Slow Steaming** is a method of reducing exhaust gas emissions. Slow steaming and just-in-time arrival are effective means to reduce fuel consumption.

There are other methods available in order to make the engines more fuel efficient. Engine manufacturers are encouraged to constantly research advanced technologies and incorporate them in order to reduce fuel consumption. Many engine manufacturers provide the ship-owners with the option to retrofit these advanced technologies on their existing vessels so that they can meet the requirements of EEXI as per MARPOL Annex VI.

In this section, we have seen the problems associated with diesel engines using fossil fuel and how they can be made more efficient in order to reduce fuel consumption. But these are part-time measures and will not result in zero-carbon shipping. Let us have a look at the ultimate solution to this problem.

7.3 SOLUTION TO THE DIESEL ENGINE CONUNDRUM

It is an established fact that the continued use of conventional fuel is not the solution to the reduction of GHG emissions and complying with the IMO strategies. The conundrum in the use of diesel engines lies in the fact that although these engines are more efficient and generate more power, they result in harmful GHGs leading to climate change. Let us now discuss some potential solutions to ensure compliance.

7.3.1 Alternate Fuels

The use of alternate fuels has long been discussed and debated. There are a number of ships sailing the oceans on alternate fuels such as methanol, ammonia, and so on. The various alternate fuels have been discussed in the previous chapter.

7.3.2 Renewable Sources of Energy

Due to the easy availability of heavy fuel oil and marine diesel oil, there was no serious thought given to shifting to renewable sources of energy for powering ships. It was only after the IMO brought about their initial strategy to reduce GHGs, followed by the final strategy, that renewable sources of energy began to be considered as a viable option for powering ships.

The main renewable sources of energy being considered are wind energy, solar energy, and electrical energy derived from renewable sources.

7.3.2.1 Wind Energy

Wind energy is a sustainable and environmentally friendly source of energy for powering ships without causing harmful emissions and pollution. Humans have been using sails to propel their boats since as early as 3500 BC. In the 17th century, the use of steam as an energy source was becoming popular in a small way. The first commercial water pump using steam was developed in 1698. More than a hundred years later, in 1807, the first commercial steamboat, known as the North River Steamboat, was developed using coal-fired steam turbines. Steamships became popular, and coal was the preferred fuel for producing steam in huge boilers. Almost another hundred years later, the first diesel engine was developed for use in ships. The rest is history, and today the vast majority of ships use diesel engines for propulsion. Wind power is being increasingly used for electricity generation across the world. Denmark is the best performer, with a share of 55% of electricity production from wind.[31]

Wind energy has many advantages, but the main advantage is that it is a renewable source of energy and has a zero-carbon footprint. There are many ways wind energy can be harnessed to power ships. Let's look at some of the common ways of doing this.

7.3.2.2 Kites

Kite sails have been around for a couple of decades on smaller yachts and boats. SkySails was the first company to develop these kite sails, and they were first tested on the ship MV Michael belonging to the Wessels Shipping Company in 2007. The 132-metre-long, 10,000-tonne vessel MS Beluga

SkySails was the first ship to be built with such a system, with a 160-square-metre kite, and launched in December 2007.

Such vessels are generally hybrid vessels with conventional diesel engines and are automated so that if the wind power is not sufficient to power the ship, the engines kick in. Pyxis Ocean,[32] the 43,291-tonne, 229-metre-long bulk carrier, was retrofitted with foldable steel and fibreglass wind sails in August 2023 and is currently in service. The vessel is owned by Mitsubishi Corporation and chartered by Cargill. Being a hybrid vessel, she can revert to normal fuel when the wind force drops. The ship is expected to reduce emissions by up to 30%. Computers monitor the wind and weather conditions and adjust the sails for maximum wind benefits.

This vessel has aroused interest in wind kite technology, and more orders are expected by the makers of the sails, SkySails. World energy leader General Electric chartered a SkySails-powered cargo vessel for transporting their cargo of equipment for power generation.

Sweden's Wallenius Marine AB, a shipbuilder and designer, in collaboration with Alfa Laval, an engineering and manufacturing giant, is in the process of building a wind-powered automobile carrier, the MV Oceanbird.

7.3.2.3 Flettner Rotor

This system of harnessing wind power uses the principle of the Magnus Effect. As per this principle, when the wind blows across the rotating sails, these sails provide a perpendicular thrust which drives the ship forward. The maximum thrust is provided in sideways winds and therefore is not effective at all times. These ships are also hybrid and have engines which automatically take over as the thrust due to the wind power reduces.

The first commercial Flettner rotor vessel, E-Ship 1, was launched in 2008. The German-flagged, general cargo vessel is 130 metre long, 12968 tonnes, and has been in service for the last 13 years. Being one of the first of its kind, the E-Ship 1 has generated a lot of interest in the shipping world and led to several orders for similar ships.

Bergebulk has made plans to install wind power on their ships. Wind sails and Flettner Rotors are being fitted on their ships in an effort to reduce their carbon footprint. Yara Marine Technologies is in the process of installing wind sails on the Berge Olympus, a 210,000 DWT bulk carrier. The London-based company Anemoi is also in the process of installing the Flettner Rotor system on some ships.

7.3.2.4 Windmill Propulsion

The concept of windmill propulsion is not new. Design and development of small vessels and yachts fitted with windmill technology have been going on for the last decade. The principle consists of a windmill which generates power to turn the propeller.

Installing windmill turbines on larger ships is a challenge by itself. Windmills have to be large to be able to generate enough power to propel larger ships. For this, several windmills need to be installed, which would not be possible on a cargo vessel. But research is on.

Archinaute, a French-based company, is at the forefront of the drive for installing windmill power on boats and yachts. The project aims to upscale the technology so that larger ships can be powered by windmills, albeit in hybrid mode, with engine power. Brittany South shipyard in Belize has constructed the Archinaute wind-powered boat, and there are several such boats on order.

Although there is a growing interest in the use of wind power for propelling ships. The concept is yet to take off for commercial reasons. *Horizon*, the European Union research and innovation magazine, has stated that there are only 21 large commercial ships across the globe which are powered by wind energy. This figure is set to double annually but is still a drop in the ocean.

Christina Aleixendri, the founder of a Spanish company called bound4blue, has been developing sail technology for ships for the past few years. Although they have not yet built very large ships, the technology is developing, resulting in regulatory compliance, cost savings, and sustainability. Recently, they signed an agreement with Amesus Shipping for installing two 17-metre-high sails on the 91-metre-long general cargo vessel M/V EEMS Traveller.[33] The installation was completed in July 2023, and the vessel is in service now.

However, wind energy usage is still in the nascent stages as a handful of ships from more than 60,000 ships sailing across the globe are propelled by wind energy.

Wind energy for ships offers sustainability, reduced emissions, and potential cost savings. By harnessing a renewable and abundant resource, ships can minimise GHG emissions and air pollution, contributing to environmental conservation. Wind power also reduces dependence on finite fossil fuels and mitigates the impact of fuel price volatility.

The drawbacks of using wind energy for ships include the following:

- There is an increased reliance on weather conditions. Variable conditions can lead to unpredictable speeds and limited range, especially during calm or adverse weather.
- Wind propulsion systems generally have lower maximum speeds compared to conventional engines, potentially affecting operational efficiency.
- Retrofitting ships with wind propulsion systems or designing new vessels to harness wind energy involves high upfront costs, which may not be acceptable to the industry.

But there is no doubt that the long-term benefits of wind energy make it an attractive option for enhancing the sustainability and efficiency of maritime transport.

Another attractive renewable energy source is solar energy which is available freely and in unlimited amounts. Let us have a look at the use of solar energy to power ships.

7.3.2.5 Solar Energy

The sun is the fundamental source of energy for the Earth. Its rays provide sustenance for habitation across the world and drive the climate and seasonal changes. However, it is only recently that solar energy is being harnessed as a source of power.

Solar power is abundantly available across the globe, but its share in electricity generation is meagre, nearly 4.6% worldwide in 2022, up from 3.7% in 2021 (Statista, July 2023).

The use of solar energy to propel ships is still in the nascent stage. Solar energy is captured using solar panels and converted to electrical power to meet the ship's energy needs. The most common method for harnessing solar energy for ships is to install solar panels on the ship's deck. These panels are made of sturdy and lightweight material which can sustain the harsh environment at sea. Another option for harnessing solar energy is to integrate solar panels into wind sails. This is especially useful on ships due to limitation of space.

Today, solar-propelled ships are being designed, which can directly power electric motors and contribute to the overall propulsion system. Solar energy is being increasingly used for powering smaller ships and yachts. But there are plans for harnessing solar energy for larger ships.

One of the first boats to run on solar power was the Solar Craft 1, launched in 1975. It achieved a speed of 1.3 knots. This was the beginning, and solar energy was subsequently incorporated into many boats and yachts. PlanetSolar was one of the largest vessels to circumnavigate the globe. Launched on 31 March 2010, the 31-metre-long catamaran had 536.65 m² of solar panels. In May 2012, the vessel completed a voyage around the globe in 584 days.

However, solar energy by itself cannot meet all of the ship's energy needs. Hybrid systems can be used instead, which combine solar energy with other sources of energy. These hybrid systems are generally automated so that during periods of sunlight, dependence on other sources of energy reduces. In addition, energy storage systems such as batteries are used to store excess energy generated during periods of high sunlight.

Wärtsilä announced the completion of their unique hybrid power system using solar power on the 199.99 metre long, 59914 DWT bulk carrier, MV Paolo Topic. The system includes an energy management system (EMS) with batteries to deliver power to the grid when solar energy is not available. Photovoltaic (PV) panels capture solar energy and transfer it to the power plant for conversion to electrical energy.

Today, many ships are operating on solar power, and many more are on order. However, considering the global fleet, solar power is yet to make a serious entry into shipping. It is hoped that in the next few years solar energy will pick up as a renewable energy source for ships, making them less carbon intensive.

Let us take a quick look at the advantages of using solar energy to power ships:

- Its renewable nature makes it an environmentally friendly energy source. It is abundantly available at no cost to the user. Once installed, solar panels offer a free source of electricity, reducing operating expenses over time. With minimal maintenance requirements and silent operation, solar systems contribute to overall operational efficiency and on-board comfort.
- It is a good alternative to the use of fossil fuels and does not produce any GHG.
- It can be used for diverse on-board applications and uses unused deck space.

However, there are also some disadvantages to using solar energy for powering ships:

- Solar panels require sufficient sunlight to generate electricity, making them less effective in regions with frequent cloud cover or during night-time operations.
- These panels are quite costly to install and require significant space, which may not be available depending on the ship type.
- Solar power systems may not be able to provide sufficient energy for high-power applications or long-distance voyages, necessitating backup power sources.
- Solar panels may get damaged in heavy weather conditions, thus requiring regular maintenance and replacement.

In conclusion, there is no doubt that solar energy is a promising option for sustainable ship operations. Using solar energy in combination with other sources of energy will mitigate some of the drawbacks. It is hoped that technological advancements will increase the range of operation and reduce the cost factor, making it an acceptable alternative to the use of fossil fuels.

Another method to reduce the carbon footprint of ships is battery-powered propulsion, which has its own advantages and disadvantages. Let us discuss this further in the next section.

7.3.2.6 Battery Power

Battery power is being increasingly considered as a viable option for reducing GHG emissions and bringing about a zero-carbon solution for shipping. However, batteries charged by on-board diesel generators are not the solution, as the carbon footprint and subsequent GHG emissions will not be reduced. Hence, it is to be ensured that the recharging is through renewable sources of energy.

Due to the constraints of storage, currently there are both fully powered and hybrid electrical systems in smaller vessels. These include tugs, ferries, and even small coastal vessels. Lithium-ion batteries are at the forefront of this transition to electrical power. They have a much longer life than other batteries and are known for their reliability. These batteries require to be charged in port. If the ship's carbon footprint is to be reduced, then the recharging should be done by the electricity obtained from renewable energy sources.

Lithium-ion batteries as a source of energy for the ship are a viable option but have several disadvantages:

- The storage capacity required could be challenging for larger ships. Large power banks are required to generate the amount of power for a larger vessel.
- Safety is a challenge where batteries are used. Damaged batteries or those exposed to heat will lead to an exothermic reaction, which will damage more batteries. This will set up a chain reaction known as thermal runaway, which may result in a serious fire. They may also release explosive gases such as hydrogen, which may result in an explosion.
- These batteries need to be recharged at specified intervals. For a ship, this may not be possible, and there is a chance of an electrical failure, which can result in a hazardous situation.

In spite of these disadvantages, the lithium-ion battery is still an attractive option for vessels. However, hybrid systems are more popular, where the battery system is in tandem with conventional diesel engines or other renewable sources of energy.

Electric propulsion will be most environmentally friendly when the electric charge is derived from renewable sources. If the batteries are charged from electric plants using conventional fossil fuels, the reduction in GHG emissions from the plant will take away the advantage of using electric propulsion. But incremental advantage will be achieved when these power plants adopt energy-efficient practices. In fact, power plants have started incorporating modern technology such as carbon capture. In this case, electric propulsion will have a greater advantage as compared to fossil fuels.

Finally, let's look briefly at tidal energy, which is perhaps the least utilised source of renewable energy. It requires further study and significant infrastructure development.

7.3.2.7 Tidal Energy

Tidal energy is the energy harnessed from tidal currents. Turbines convert the kinetic energy from the tidal flow to electrical energy. It is a renewable source of energy and is a zero-carbon option. The consistent and predictable nature of tidal patterns makes it a reliable power source. Further, it is a renewable source of energy without any GHG emissions. But the use of tidal energy also comes with disadvantages such as high investment cost, possible environmental impacts, and geographical limitations.

Tidal energy is, therefore, a promising source of energy, but it comes with its own set of disadvantages. It is up to the different administrations to deal with the disadvantages and encourage the installation of tidal power plants to harness this freely available and sustainable source of energy.

Some ports are in the process of installing such turbines to meet their electrical needs. Excess energy can be transferred to the main electrical grid. Ships that require shore energy can be plugged into these renewable sources when alongside ports; this is also known as "cold ironing". Some ships that continuously operate in tidal waters can be fitted with tidal turbines, which can convert the tidal energy to electrical energy.

As we approach the deadlines for reducing global warming by controlling GHG emissions, ship-owners are taking proactive action. New ships that can use alternative fuels are being ordered, Ship Energy Efficiency Management Plan (SEEMP) is being strictly followed to reduce the carbon footprint of the ships, and both ship and shore staff are being made aware of the dangers of global warming and the precautions that are to be taken to control it. Renewable sources of energy are increasingly being used in order to reduce the use of fossil fuels. There are many wind- and solar-powered ships that are in the final stages of design and development. As building technology advances and construction of such ships become more cost-effective, they could well become popular as a means of reducing the carbon footprint of ships.

IMO and other regulatory bodies have been focusing on reducing GHG emissions from ships for more than a decade. Conventional diesel engines have proved to be emitters of such gases and contributing to global warming. There is no doubt that the solution lies in switching to alternate fuels and renewable energies. But incorporating these measures in the shipping industry is not an easy task, mainly because they are not easily available and not cost-effective. However, it is expected that with new technological developments, alternate fuels will become cost-effective and will be available across the globe. Until such a time, the introduction of new technology will result in reducing fuel consumption and emissions.

7.4 SUMMARY

In this chapter, we have discussed the advantages of using conventional fossil fuels for ships' engines. But as we are aware, the major disadvantage of using fossil fuels is GHG emission, resulting in global warming. Technological developments, including exhaust gas recirculation (EGR), selective catalytic reduction (SCR), organic Rankine systems (ORC), are helping to increase fuel efficiency of engines. We also present some alternate solutions in this chapter, for example, renewable sources of energy, such as wind energy, solar energy, and so on. This will help the reader to understand how ships can reduce their dependence on fossil fuels and move onto to alternative fuels and renewable sources of energy.

In the next chapter, we shall discuss how energy audits can help to increase the fuel efficiency of ships, as well as how to conduct such audits.

BIBLIOGRAPHY

30. DNV Maritime Forecast to 2050. (2024). https://www.dnv.com/maritime/pub
    lications/maritime-forecast-2023/index.html.
    This report by DNV gives the maritime forecast to 2050 covering the IMO
    strategy for GHG reduction as well as the integrated approach to decarbonisation.
31. Data Page: Share of Electricity Generated by Wind Power, part of the following
    publication: Hannah Ritchie, Pablo Rosado and Max Roser. (2023). *Energy*.
    Data adapted from Ember, Energy Institute. https://ourworldindata.org/gra
    pher/share-electricity-wind [online resource].
    All relevant data related to wind production from across the world is available in this report.
32. The Wind-Powered Pyxis Ocean. (13 March 2024). https://www.cargill.
    com/2024/first-wind-powered-ocean-vessel-maiden-voyage.
    Cargill revealed the test results of the wind powered ship. The test results
    were satisfactory and within expectations.
33. Christina Aleixendri. (4 July 2023). *Wind Powered Ships by Bound4blue*.
    https://bound4blue.com/the-general-cargo-vessel-eems-traveller-ready-to-set-
    sail-today-with-two-suction-esails/.
    This company specialises in wind technology and is one of the pioneers in
    wind-powered ships.

Chapter 8

Energy Audits, Surveys, and Certification

Energy efficiency and prevention of air pollution from ships is a mandatory requirements specified in MARPOL Annex VI, applicable to all vessels of 400 gross tonnage and above. In order to ensure compliance with these regulations, ships have to obtain relevant certifications.

In this chapter, we will discuss the certification system to ensure that the ships are in compliance with International Maritime Organization (IMO) regulations, as well as the statutory requirements of their flag administration. We will also see the survey procedures leading to the issuance of these certificates. This is important because these certificates are mandatory requirements and ships are liable to be detained or barred from entering the port in their absence of these certificates. In addition, we will discuss the energy audits that a ship-owner or operator may arrange to conduct on board. Such audits are voluntary and help both the ship staff and the management personnel ashore to identify shortfalls in the system and rectify them. Audits are important because a ship that has been audited can take corrective and preventive actions in order to rectify the non-conformities detected. This will help the ship to be in continuous compliance with the regulatory requirements and ensure that the statutory surveys are successfully completed.

These audits and surveys help in improving energy efficiency and reducing greenhouse gas (GHG) emissions by identifying inefficiencies, recommending improvements, and ensuring compliance with energy efficiency standards. Ship operators are increasingly resorting to voluntary audits to identify shortfalls in energy efficiency systems and to take corrective as well as preventive actions.

The International Air Pollution Certificate is a statutory certificate issued to those ships found to be in compliance with Annex VI of MARPOL. Let us have a look at the procedures for issuance of this certificate.

8.1 INTERNATIONAL AIR POLLUTION PREVENTION CERTIFICATE

This certificate confirms that the ship is complying with the requirements of Annex VI of MARPOL related to air pollution. During this survey, the documentation related to the SO_x and NO_x emissions, VOC management plan, and other documents as evidence of compliance with Annex VI. The equipment is also inspected for proper operation, and if it is in compliance, the IAPP certificate is issued by the flag administration of the vessel. This certificate is valid for five years.

These surveys are usually carried out by the RO on behalf of the Administration. To ensure continuous compliance, annual surveys and endorsements, as well as intermediate surveys, have to be successfully conducted.

The initial survey is conducted before the ship is put into service. These surveys are generally conducted by the flag administrations or by their approved RO on their behalf. The surveyors inspect the ship, her equipment, as well as the documents to confirm that she is constructed, equipped, and maintained as required by MARPOL Annex VI.

The renewal survey is conducted to ensure that the ship continues to be in compliance with the MARPOL Annex VI requirements. These surveys can be conducted three months before the end of the validity period of the certificate onwards. The IAPP certificate is valid for five years, and before the expiry of this certificate, the survey has to be completed satisfactorily so that the vessel is always in possession of a valid certificate. In case the ship is not in a port where she is to be surveyed, the Administration may grant an extension of three months so that the ship can complete her voyage to the port of survey.

During the initial and renewal surveys, it will be confirmed that the ship complies with the requirements of Annex VI of MARPOL as follows:

- Regulation 12: Ozone Depleting Substances (ODS): Details of ODS on board, along with their recharge, repair, and maintenance. The ODS record book will be verified, with details of deliberate or accidental discharge of ODS to the atmosphere, to land-based facilities, as well as supply of ODS to the ship.
- Regulation 13: Nitrogen Oxides (NO_x): The Engine International Oil Pollution Certificate (EIAPP) as well as the approved technical file is available on board, and all engines are certified, and the ship possesses the necessary documents. The record book of engine parameters is sighted and verified that it is updated.
- Regulation 14: Sulphur Oxides (SOx): The sulphur content in the fuel used on board in both ECA areas (0.1% m/m) and outside ECA areas (0.5% m/m) is verified. The system of changeover from high to low sulphur fuels is verified for compliance with the requirements. The operation of the exhaust gas cleaning system, where fitted, is verified.
- Regulation 15: Volatile Organic Compounds (VOC): This is valid for tankers when entering or inside ports. Vapour return systems and their certification are verified. The condition of the vapour collection system, including drains, valves, flanges, testing systems, and alarm systems, is verified.

- Regulation 16: Shipboard Incineration: The certificate and operation manual are checked to confirm that the equipment is approved. The equipment is checked for satisfactory operation and confirmed that instructions for operation have been posted. Drip trays should be available where required, and a list of materials prohibited from incinerating is posted in the engine room to ensure all staff are aware of these restrictions.
- Regulation 18: Fuel Oil Quality: This includes the verification of bunker delivery notes (BDNs) for all bunker operations and fuel oil samples. As per IMO requirement, a sample of the fuel being delivered is taken during the bunkering process. This is known as the MARPOL sample, and it is to be confirmed that every BDN is accompanied by this sample of a minimum volume of 400 ml. The BDN is stored onboard for three years along with the samples.

The intermediate survey is conducted within three months before or after either the second or third anniversary date of the certificate. The objective of the intermediate survey is to ensure continuous compliance with the requirements of MARPOL Annex VI. The IAPP certificate is endorsed with the date of satisfactory completion of the intermediate survey.

The annual survey is conducted within three months before or after every anniversary date of the certificate. The survey is conducted to ensure that the equipment and arrangements comply with the requirements. The annual survey is also endorsed on the certificate.

The IAPP certification is evidence that the ship is complying with the requirements of MARPOL Annex VI in general. Now that we're aware of its requirements, let's examine the Engine International Air Pollution Certificate (EIAPP), which relates specifically to the ship's engines and that they conform to the emission requirements for NO_x.

8.2 ENGINE INTERNATIONAL AIR POLLUTION PREVENTION CERTIFICATE

This certificate confirms that the main engine of the ship meets the NO_x emission standards as set out in MARPOL Annex VI. It is required for engines with a power output greater than 130 kilowatts. It is issued to vessels who have successfully completed a survey and testing of NO_x emissions as per the NO_x technical code. The certificate has lifetime validity, but the ship should keep records related to the engine's performance throughout the period. Any modifications to the engine that may affect its NO_x emissions will need to be recorded, and the EIAPP certificate may be revised accordingly. The certificate is issued by the engine manufacturers and comes with a supporting technical file containing details on the engine components, allowable engine adjustments, and on-board NO_x verification procedures.

During the survey, the emission levels are checked in order to ensure that they meet the IMO requirements as specified in MARPOL. It is confirmed that the emission monitoring system is installed and operating satisfactorily so that emissions can be checked regularly. The maintenance records are checked to confirm compliance with manufacturers' recommendations and regulatory requirements.

The certificate is issued by the flag administration of the ship after verifying compliance with the applicable requirements. It serves as evidence that the ship's engines meet the necessary emission standards, allowing the vessel to operate in international waters without violating air pollution regulations. Failure to comply with EIAPP requirements can result in penalties and sanctions by the port state control and other administrative agencies.

The EIAPP certificate directly addresses the issue of pollution prevention from ships. Its requirements and the measures taken to comply with them can lead to indirect improvements in the energy efficiency of ships. By promoting the adoption of cleaner technologies, alternative fuels, optimised operations, and innovation in the maritime industry, the EIAPP contributes to more sustainable and efficient operations on board.

Another important certification is the International Energy Efficiency Certificate (IEEC), which relates to the energy efficiency of the ship and that the ship is in compliance with the MARPOL regulations regarding energy efficiency. Let's look at this next.

8.3 INTERNATIONAL ENERGY EFFICIENCY CERTIFICATE

Chapter 4 of MARPOL Annex VI deals with the relevant energy efficiency regulations. Every ship of 400 gross tonnes and above engaged in international waters shall be in possession of the IEEC issued by the Administration or their ROs. The certificate has a lifelong validity and should be supported by the record of construction for ship energy efficiency, Ship Energy Efficiency Management Plan (SEEMP), Statement of Compliance for DCS, and the EEDI technical file. The certificate will be issued after an initial survey of the new ship provided the SEEMP is on board and the EEDI has been verified. In case of a major conversion of the ship affecting its energy efficiency, a general or partial survey is to be carried out to revise the IEEC. Since the certificate is issued by the administration, when the ship is transferred to another flag, a new certificate will have to be issued after a further survey.

The IEEC certificate is issued to ships during the initial survey after verifying that the EEDI meets the requirements and the SEEMP is being implemented on board. In case of a major conversion which affects the engines or the energy efficiency of the ship, the IEEC is to be re-issued.

The IEEC must be accompanied by a Statement of Compliance. This statement is issued by the Administration and states that the ship has submitted the data required by Regulation 22A of MARPOL Annex VI and that the data was collected and reported as per the methodology and processes set out in the ship's SEEMP. It is valid for the year that the statement is issued and for the first five months of the following year.

IMO Resolution MEPC 278(70),[34] adopted on 28 October 2016, details the amendments to MARPOL Annex VI related to the DCS for fuel oil consumption of ships. The methodology for collecting such data shall be included in the SEEMP.

Now that we're aware of the various certifications that are required, it is important to turn our attention to energy audits.

8.4 INTRODUCING ENERGY AUDITS

An energy audit is the first step towards adopting an exercise in energy efficiency on ships and harvesting the resultant savings. Although energy audits are not a mandatory requirement for ships, they are crucial for promoting energy efficiency and sustainability on ships. They should be conducted regularly to ensure ongoing improvements in energy efficiency, cost reduction, and environmental performance. Collaboration with experienced consultants and energy management experts is essential for a successful audit and implementation of energy-saving measures.

Energy audits on board ships are especially relevant in the current scenario where the IMO is breathing down ship-owners' necks with stringent requirements of fuel efficiency, carbon footprint, and so on. In this section, we will look at why energy audits are necessary and how they are conducted.

8.4.1 Understanding the Need for Energy Audits

Ship-owners need to be aware of the shortcomings in energy efficiency on board and the corrective actions required. Energy audits can help address this and are important for the following reasons:

- **Preparing the Vessels for the Statutory Surveys:** The IAPP certificate requires an initial survey followed by annual surveys and one intermediate survey. Before the expiry date of the survey, a renewal survey has to be successfully completed. The EIAPP certificate and the IEE certificate require an initial survey as well as a follow-up survey in case of major modifications or changes of flag administration. In order to ensure a smooth survey without any hiccups, many ship-owners resort to a "gap audit" to identify the gaps in the ship's preparation for the survey so that corrective action can be taken.
- **Cost Benefit in Ship Operations:** An energy audit can identify the energy losses on board and focus on cost-effective measures to improve energy efficiency on board. Energy costs are a major percentage of the

ship's operational costs, and any reduction in this would directly relate to a cost benefit for the ship-owner or operator.

- **Compliance with ISO 14001 and 50001 Requirements:** Many shipping companies implement ISO 14001 (environmental management systems) and ISO 50001 (energy management) for furthering their business or simply for ethical considerations. Both these ISO standards require energy and environmental audits on ships and in shore offices. In addition, compliance with regulatory requirements, as well as maintaining a good CII score, will go a long way to enhance the image of the ship and company in the competitive commercial markets.
- **Minimising Chances of Deficiencies and Delays by Port State Control and Other Regulators:** Regulatory bodies such as port state controls and even port authorities routinely board vessels to look for shortfalls in compliance with MARPOL Annex VI requirements. Energy audits would pinpoint these deficiencies beforehand and protect the ship from fines and deficiencies.
- **For Ethical Considerations:** Many ship-owners are sincere in their belief that GHG emissions need to be reduced in order to control global warming. Energy audits are the solution to achieve their goals and reduce the carbon footprint of their ships.
- **Evaluating New Technologies:** Energy audits are a means to assess and recommend the adoption of new technologies that can enhance energy efficiency. By knowing the shortcomings in the system, the ship-owners are made aware of the need to implement better practices, optimise operations, and, more importantly, incorporate new technologies to increase fuel efficiency. This might involve upgrading equipment, implementing more efficient propulsion systems, or adopting renewable energy sources such as solar panels or wind turbines. These technological advancements can lead to significant cost savings over time.

Energy audits are essential for improving energy efficiency by detecting energy losses and inefficiencies, assessing equipment performance, and recommending efficiency measures. These audits enable ship operators to identify sources of energy waste and opportunities for upgrades or retrofits. They can assess the potential savings in terms of energy consumption, costs, and environmental impact. Implementing the recommended measures can result in financial savings, reduce GHG emissions, and enhance operational sustainability.

Let us now have a look at the objectives and the methods of conducting energy audits.

8.4.2 Objectives of Energy Audits

Energy audits are conducted to help the ship-owners optimise energy use, comply with regulatory requirements, and enhance environmental sustainability in the shipping industry. In general, they improve energy efficiency as well as the ship's performance. However, let's take an in-depth look at their key objectives:

- **Identifying Energy Loss and Inefficiencies:** The main objective of an energy audit on board a ship is to identify the areas on the ship from where energy losses can occur. This includes leaks, equipment malfunction, and other system inefficiencies. Some methods for achieving this are as follows:
  - Comprehensive data on ships' energy consumption is collected from all sources. On-board inspections are conducted by the auditors to look for leaks, wear and tear, pipes, and other potential areas of energy loss.
  - Auditors use performance analysis tools to assess the performance of various machinery such as engines, generators, mooring gear, and other operating machines. Energy consumption patterns are identified by analysing data trends during different operational conditions. From these patterns, irregularities in energy consumption can be identified, and corrective actions can be suggested accordingly.
  - Manufacturers often conduct energy audits on demand by the ship-owners of their engines. They identify thermal irregularities using thermal imaging techniques. Uneven temperature distribution or hotspot areas may indicate worn-out bearings, insulation issues, and so on. This leads to energy loss and reduces the efficiency of the related machines.
  - Excessive energy consumption can be identified by installing energy meters and other monitoring devices on engines and other critical systems. Energy usage patterns can then be used to identify excessive consumption of energy.
  - Auditors conduct analyses on the ship's exhaust gas to understand its composition and thereby identify possible fuel inefficiencies leading to increased fuel consumption.
- **Emission Reduction:** By identifying fuel inefficiencies, energy audits play a vital role in reducing emissions. Ship-owners, operators, and ship staff can take corrective action based on the audit report to plug energy leaks. Combustion processes can be optimised, maintenance schedules can be taken up, replacement of identified worn-out parts can be undertaken, and the overall energy efficiency of the engines can be improved.
- **Address Operational Inefficiencies:** Energy-efficient operations are the pillars of ship energy efficiency. Although SEEMP provides a proper structure for operational optimisation, audits can help identify whether operational requirements are being followed. Ensuring that there is no operational waste of energy will result in reducing fuel consumption and increasing the energy efficiency of the ship. Auditors check the records, inspect the accommodation and other spaces, and

evaluate the functional usage of energy to identify operational energy loss. Once identified, corrective and preventive actions can be taken.

- **Ensure Regulatory Compliance:** One of the primary objectives of energy audits is to check for regulatory compliance. With the IMO having brought in strict requirements regarding EEDI, EEXI, CII, and so on, auditors crunch the vessels' data to confirm compliance. This is done as follows:

  - **EEDI:** For new vessels, the auditors check the vessel's technical file to confirm that the EEDI mentioned is in line with the IMO requirements. This file contains the EEDI calculations carried out by the shipyard and verified by the classification society. The IMO Resolution MEPC 365(79)[35] enumerates the preliminary verification of the attained EEDI to be carried out at the design stage, the final verification at the sea trials, and the verification required in case of major conversion.

  - **EEXI:** Existing vessels are required to have on board an approved EEXI technical file. This is generally verified during the IAPP survey. Energy auditors verify the file and confirm compliance with the requirements. If vessels are not found to be in compliance, the auditor will issue a non-conformity report. In such a case, the vessels have to take corrective actions to reduce their GHG emissions with the help of their classification society. Energy improvement measures will have to be initiated to ensure that the vessel is in compliance with the IMO EEXI requirements.

  - **CII:** IMO has introduced the CII, effective 1 January 2023 in order to measure the carbon footprint of the ships. The CII measures the vessel's carbon intensity over a period of one year. It is to be confirmed during the audit that the vessel is in compliance with the DCS and has on board the Statement of Compliance, which is a mandatory requirement. Based on the ship's CII, it will be rated A, B, C, D or E, as discussed in Chapter 3, which will be recorded in the Statement of Compliance.

Now let us have a look at the key elements of energy audits. This will help us to understand the procedure and process of conducting energy audits.

8.4.3 Key Elements of Energy Audits

As discussed, an energy audit is a comprehensive examination that uncovers crucial information often overlooked or taken for granted. It delves into the ship's energy consumption both at sea and in port on a daily basis. The key elements typically included in an energy audit are as follows:

1 **Energy Consumption Analysis:** The audit begins by meticulously analysing the ship's energy consumption patterns, capturing data for both

sea voyages and port stays. This data provides a baseline for assessing current energy usage.

2 **Process Identification:** It identifies the various processes and operations on the ship that consume energy. This includes propulsion, lighting, heating, cooling, electrical systems, and auxiliary machinery.

3 **Energy Efficiency Assessment:** The audit evaluates the efficiency of these energy-consuming processes and identifies areas where energy is being wasted due to inefficiencies or suboptimal operation. It looks for opportunities to optimise these processes.

4 **Energy Saving Recommendations:** Based on the assessment, the audit generates a list of recommendations for achieving energy savings. This may include suggestions for equipment upgrades, procedural changes, or the adoption of new technologies.

5 **Cost-Benefit Analysis:** The audit quantifies the potential cost savings associated with implementing the recommended changes. It helps ship operators understand the financial implications of energy efficiency improvements.

6 **Management Control Evaluation:** An essential aspect of the audit is assessing the effectiveness of management in implementing the recommended energy-saving measures. This involves examining whether the necessary policies, procedures, and training are in place to facilitate these changes.

7 **Environmental Impact:** The audit considers the ship's environmental impact, especially in terms of emissions and carbon footprint reduction. It aligns energy-saving efforts with broader sustainability goals and global warming mitigation.

8 **Compliance and Regulations:** It ensures that the ship complies with relevant energy efficiency regulations and international standards, helping ship operators avoid penalties and maintain legal compliance.

9 **Monitoring and Continuous Improvement:** An effective energy audit includes a plan for ongoing monitoring of energy consumption and the impact of implemented changes. This allows for continuous improvement in energy efficiency over time.

In summary, an energy audit is a comprehensive assessment of a ship's energy usage, processes, and potential for energy savings. It provides ship operators with valuable insights, recommendations, and a roadmap for improving energy efficiency, reducing operational costs, and contributing to environmental sustainability.

Let us now have a look at the practical process of conducting an energy audit on ships. The audit should be conducted without any blame game and in a collaborative manner.

8.4.4 Understanding Energy Audits

Energy audits for ships are essential for improving energy efficiency, reducing operational costs, and minimising environmental impact. Let's first look at the process of conducting such an audit, followed by learning about the auditors who are qualified to conduct them.

8.4.4.1 Conducting an Energy Audit

The typical processes involved in conducting energy audits for ships are as follows:

1. **Opening Meeting:** The audit begins with an opening meeting attended by the auditors and the auditees. The lead auditor addresses the meeting and makes it clear that all parties to the audit are working towards a common goal of identifying system shortfalls and putting in place corrective and preventive actions so as to achieve the common game of energy efficiency.
2. **Familiarisation:** The auditors study the ship's SEEMP as well as the Safety Management System in order to get a correct estimate of the ship's regulatory requirements. A random sampling of the ship's compliance with the SEEMP requirements should be carried out. This may include documentary evidence of compliance as well as the ship's operations.
3. **Data Collection:** The auditors go through the ship's record and gather data on the ship's fuel consumption, engine performance data, voyage records, ship's operational parameters, loading conditions, and so on.
4. **Measurement of Energy Consumption:** The energy consumption of shipboard systems, such as engines, generators, auxiliary equipment, lighting, air conditioning, and refrigeration systems, is evaluated. It is important for the auditors to document the fuel consumption and energy use during various operational modes (e.g., cruising, idling, and port operations).
5. **On-Board Inspection:** A physical inspection of the ship to identify energy-saving opportunities and assess the condition of equipment and systems is to be conducted in order to identify any shortfalls. This includes the following:
   - Inspection of insulation, seals, and other systems is to be carried out to identify leaks and inefficiencies.
   - The condition of propulsion and power generation equipment is to be checked in order to confirm that the equipment is well maintained and working satisfactorily without any wastage of energy.
   - Review the ship's design specifications and maintenance records such as the planned maintenance system (PMS) to ensure that maintenance is being carried out routinely as per the maker's instructions and industry standards.

- The audit checklist issued by the auditing organisation is to be followed. This will ensure that nothing of importance is missed.

6. **Data Analysis:** The collected data is closely examined in order to identify patterns and areas of inefficiency. An energy balance for the ship is created, if not already available on board, so that energy flows and losses can be tracked. Auditors may use specialised software tools or models to simulate various scenarios and assess potential energy-saving measures.

7. **Interviews:** The ship staff are interviewed to find their level of application of energy-saving methods on board. Familiarisation with SEEMP and the crew's knowledge of individual roles and responsibilities are confirmed. Crew competence is not tested; the objective is to find out what training, if any, is to be imparted to the crew, whether the senior officers are motivating the crew towards energy efficiency, and whether procedures are in place to define and document these measures.

8. **Report:** One of the important elements of the energy audit is to prepare a factual report of the findings with details of energy consumption across various systems, specifying where wastage of energy is occurring and the possible reasons for the same. The report should also specify corrective actions to plug the energy wastage and preventive actions to avoid repetitions of the non-conformities observed. Non-conformities that are observed are to be mentioned clearly, along with the corrective actions required and the timeframe for this. The report should be made in a comprehensive manner in order to project the correct situation of energy efficiency on board the ship without any bias or partiality.

9. **Closing** Meeting: This is conducted after the audit and the preparation of the reports. The reports are discussed during the closing meeting with special emphasis on the non-conformities observed. The corrective as well as preventive actions are highlighted along with training requirements or changes in the ship's energy efficiency strategies so as to reduce GHG emissions by reducing energy losses and increasing energy efficiency.

These audits can be quite rigorous and last for two to three days. The identification of non-conformities is a vital aspect of energy audits as it helps identify areas and practices which require improvement so that energy efficiency is enhanced. If the auditors observe that the ship is not in compliance with the requirements, they will issue non-conformities to the ship which has to be corrected within a given period. It is necessary for the ship staff to cooperate with the auditors and take a positive view of the auditing process. Only then can a holistic view of the ship's energy efficiency efforts be obtained and improvements identified.

Let us now have a look at the auditors qualified to conduct energy audits on board ships.

8.4.4.2 Qualified Auditors

They typically include professionals with backgrounds in marine engineering, naval architecture, and energy management. These experts possess knowledge of marine propulsion systems, hull design, and energy efficiency measures applicable to maritime operations. They may work for classification societies, certification bodies, or as independent consultants. They play an important role in improving the energy performance of ships, reducing fuel consumption, and minimising the environmental impact.

These auditors must also be certified for conducting such audits. This involves a combination of education, training, professional experience, and successful completion of examination or assessment criteria. These certifications help ensure the competence and credibility of energy auditors, promoting confidence in the audit process and the effectiveness of energy efficiency measures leading to sustainable operations in the maritime industry.

In conclusion, we see that energy audits, surveys, and certification programmes are essential tools for improving the energy efficiency of ships and reducing their carbon footprint. By identifying opportunities for optimisation, implementing efficiency measures, and ensuring compliance with regulations, they contribute to the global effort to mitigate climate change and achieve a more sustainable maritime industry.

8.5 SUMMARY

In this chapter, we have seen the necessity and importance of energy audits. As we have discussed, these audits are important on board ships as they help identify deficiencies in the ships' energy efficiency. In addition to discussing the various certifications related to environmental protection, this chapter also explores the reasons why it is important to conduct an energy audit. The key elements of an energy audit are then detailed, along with the process of conducting such audits. Further, the IAPP certificate along with the necessary surveys for the same are elaborated. The practical aspects of conducting audits on board the ship are also discussed.

In the next chapter, we shall discuss the international legislation regarding emissions, which are generally long-winded and difficult to comprehend. The next chapter explains these in simple terms.

BIBLIOGRAPHY

34. IMO Resolution MEPC 278(70). (2016). *Data Collection System*. https://wwwcdn. imo.org/localresources/en/OurWork/Environment/Documents/278%2870% 29.pdf.

This resolution states the requirements for collecting and reporting of ship's fuel oil consumption data as well as the Statement of Compliance.

35. IMO Resolution MEPC 365(79). *2022 Guidelines on Survey and Certification of the Energy Efficiency Design Index (EEDI)*. https://wwwcdn.imo.org/local resources/en/KnowledgeCentre/IndexofIMOResolutions/MEPCDocuments/ MEPC.365(79).pdf.

This resolution gives the procedure for verification of EEDI during the survey and certification.

Chapter 9

International Legislation Regarding Emissions

The effect of GHGs and global warming has been the subject of study and debate for a long time. One of the earliest studies undertaken by the American Institute of Physics[36] has compiled a timeline of global warming. The following are some of the milestones highlighted in their report (from 1800 to 1970):

- 1800–1870: The levels of CO_2 in the atmosphere, as later measured in ancient ice, were about 290 ppm (parts per million). The mean global temperature was roughly 13.6° C
- 1896: Arrhenius published the first calculation of global warming from human emission of CO_2.
- 1920–1925: Opening of Texas and Persian Gulf oil fields inaugurates an era of cheap energy (fossil fuel)
- 1968: Studies suggest a possibility of collapse of Antarctic ice sheets, which could raise sea levels catastrophically.
- 1970: First Earth Day. Concerns spread about global degradation. This led to the creation of the United States Environmental Protection Agency (USEPA), the Clean Air Act, as well as the Clean Water Act.

According to the report, as of 2022, the "mean global temperature is 14.8°C, the warmest in tens of thousands of years. Level of CO_2 in the atmosphere is 421 ppm, the highest in millions of years". This is an interesting and alarming statement, which is the reason that every effort is being taken by all concerned to reduce global warming. Sustainable development requires that while the world moves ahead in various fields, the environment and sustainability are taken care of.

In this chapter, we will discuss the various initiatives and efforts taken across the globe over the last many years to address climate change. The shipping industry is part of the global industry; to understand the regulatory requirements on ships, we need to know the global scenario and the worldwide focus on GHG emissions.

Let us have a look at the various summits, conferences, and protocols held by the global stakeholders to address climate change and resulting global warming.

DOI: 10.1201/9781032702568-9

9.1 GLOBAL INITIATIVES TO ADDRESS CLIMATE CHANGE

One of the first to study and conclude that global temperatures had risen was Guy Callendar, who analysed temperature data in the 1930s to reach his conclusion. The famous Mauna Loa Observatory measurements conducted by Charles David Keeling showed an increase in carbon dioxide levels in the Earth's atmosphere.

Soon the spectre of increased carbon dioxide levels and the resultant global warming became clear to all the stakeholders, and the world started to consider ways and means to combat climate change. This resulted in many innovative and practical initiatives and solutions which were implemented with the passage of time. These global initiatives helped mobilise political will, foster cooperation, and encourage action to address climate change. By leveraging collective strengths and resources, these initiatives contribute to building a more sustainable world for current and future generations.

The first global conference on the environment, the United Nations Conference on Environment and Development (UNCED), commonly known as the Stockholm Summit or the United Nations Conference on the Human Environment, was organised by the United Nations in Stockholm, Sweden between 6 and 15 June 1972. It resulted in the Stockholm Declaration,[37] consisting of 26 principles dealing with various aspects related to environmental issues. It motivated countries across the world to consider the environment and create ministries and agencies for monitoring and protecting the environment. It also created the United Nations Environment Programme (UNEP), focusing solely on environmental issues.

The Stockholm Summit was a landmark event in the history of environmentalism and sustainability. Although it did not lead to any immediate binding agreements or treaties, it laid the foundation for global environmental governance and raised awareness about the urgent need for sustainable development.

In 1979, the United Nations Economic Commission for Europe (UNECE) adopted the Convention on Long-Range Transboundary Air Pollution (CLRTAP) to address the air pollution that crosses national borders. The convention aimed to protect the environment against air pollution, and it entered into force in March 1983.

The Montreal Protocol on Substances that Deplete the Ozone Layer, finalised in 1987, is a global agreement in order to prevent the depletion of the ozone layer. The ozone layer prevents harmful ultraviolet rays from passing through the Earth's atmosphere. Ozone depletion will lead to severe skin diseases, affect agricultural production, and disrupt the marine ecosystem. The Montreal Protocol focused global attention on the importance of the ozone layer, and its implementation will result in the recovery of the depleted ozone layer by the middle of the 21st century.

The Montreal Protocol was primarily aimed at protecting the ozone layer, but its provisions and success indirectly contributed to mitigating GHG emissions, thereby aiding in the global efforts to combat climate change.

The Intergovernmental Panel on Climate Change (IPCC) was established in 1988 by the United Nations under the auspices of the UNEP and the World Meteorological Organization (WMO). The IPCC has the basic objective of studying the different aspects of climate change and producing reports and guidelines to enable policymakers and governments to formulate policies and intergovernmental agreements. Their assessments are globally accepted and have contributed to efforts to address climate change.

IPCC resulted in directing global attention towards mitigating global warming through scientific assessments. Policymakers, governments, and the public were thus informed about the risks and impacts of climate change. The IPCC also provided guidance to international efforts to mitigate climate change, thereby contributing to the collective global effort to address this urgent challenge.

The United Nations Framework Convention on Climate Change (UNFCCC) was adopted in May 1991 during the First Earth Summit in Rio de Janeiro, Brazil, and entered into force in March 1994. The main objective of the convention is to prevent human interference with climate change leading to a concentration of GHGs and subsequent degradation of human health and global warming.

UNFCCC is an important international treaty aimed at combating climate change by fostering global cooperation and coordination among nations. The convention sets targets and commitments, mobilises resources, and facilitates action to address climate change globally.

One of the important outcomes was establishing mechanisms such as the Green Climate Fund (GCF) to provide financial assistance to developing countries, helping them to transition to low-carbon and climate-resilient economies.

The First Earth Summit or the United Nations Conference on Environment and Development (UNCED) was held in Rio de Janeiro, Brazil from 3 to 14 June 1992, on the occasion of the 20th anniversary of the Stockholm Summit held at Stockholm in 1972. The summit was attended by political leaders, diplomats, scientists, non-governmental organisations (NGOs), and media personnel from 179 countries. The main objective of the Rio Earth Summit was to establish an agenda and a fresh blueprint for action on environmental issues and sustainable development. Agenda 21,[38] released during the summit, addresses various sectors and encourages countries to develop strategies for sustainable development in the 21st century. Agenda 21 outlined various strategies for addressing environmental degradation, poverty, and social inequality while promoting economic growth and prosperity.

The Rio Summit established the foundation for tackling climate change and promoting sustainable development. The summit resulted in the Rio Declaration on Environment and Development, which established 27 principles for sustainable development and the right to development. Most importantly, it introduced the principle of "the polluter pays", which states that those who produce pollution or degrade the environment should bear the costs associated with their actions. This was important as it made the polluter accountable and encouraged others to adopt cleaner practices. These principles provided a guiding framework for future environmental and development policies.

The Kyoto Protocol is an extension of the UNFCC and was adopted in December 1997 in Kyoto, Japan. It entered into force on 16 February 2005. It set binding targets for developed countries to reduce harmful emissions. The Kyoto Protocol defined these emissions as carbon dioxide, nitrous oxide, methane, hydrofluorocarbons, sulphur hexafluoride, and perfluorocarbons. The emissions of these GHGs were required to be reduced below 1990 levels during the commitment period from 2008 to 2012. Market-based mechanisms were introduced to ensure compliance, such as emissions trading (ET), where countries emitting less than their allocated limits can sell their allowances to those countries emitting more than their limits. It encouraged the provision of technical and financial assistance to developing countries in order to help them meet their targets.

Unfortunately, the Kyoto Protocol was considered a failure because the United States did not agree to the provisions and did not sign up. Further, some large economies like the People's Republic of China were excluded from any lasting commitment because they were classified as developing countries and wished to concentrate more on economic growth and poverty alleviation. Under the UNFCC principle of "common but differentiated responsibilities", developing countries were exempt from mandatory emission reduction commitments. Developed countries viewed it as a costly punishment and did not agree to this, due to which it failed to ignite the interest of the parties and had little impact on environmental sustainability.

Eventually, the Paris Agreement was initiated to address the shortcomings of the Kyoto Protocol.

Twenty years after the Earth Summit, the United Nations Conference on Sustainable Development (UNCSD), commonly known as Rio+20, was held in Rio de Janeiro, Brazil, in June 2012. The aim of this summit was to foster sustainable development and environmental protection and achieve economic growth while addressing poverty eradication. World leaders, non-governmental organisations (NGOs), business entities, as well as members of the civil society were invited to meet and find solutions to important environmental and developmental issues. UNCSD helped tackle climate change by reaffirming international commitments, promoting green economy initiatives, and strengthening institutional frameworks. By advancing the development of sustainable development goals (SDGs) and engaging stakeholders in collaborative efforts, the conference helped address environmental challenges of sustainable development.

One of the main outcomes of the UNCSD was the release of the document *The Future We Want*.[39] It embodies a shared vision and dedication from the global community to promote sustainable development worldwide.

The Paris Agreement[40] can be considered to be a sequel to the Kyoto Protocol. It brought about definitive changes in the world's reaction to global warming. It was adopted in Paris, France, in December 2015, and it entered into force on 4 November 2016. The objective of the agreement was to bind the countries to limit their temperature increase to within 2°C above preindustrial limits and pursue efforts to limit it to within 1.5°C. The Paris Agreement works on a five-year cycle of increasing climate action and introduced the concept of nationally determined contributions (NDCs). Each successive NDC should reflect a higher degree of ambition and action compared to the previous NDC. The Paris Agreement has been considered a success as compared to the Kyoto Protocol because due to its broader inclusivity, garnering commitments from a greater array of nations, including significant emitters such as China and the United States. Countries like China, Japan, and the European Union were encouraged to set carbon-neutral goals. Overall, 195 parties signed the agreement and promised to take action against global warming.

The Paris Agreement provided a framework for collective action, promoted transparency and accountability, supported developing countries, and set the world on a path towards a more sustainable and resilient future. Unfortunately, other than monitoring and reporting carbon emissions, the agreement did not have any coercive power to force any country to reduce their emissions.

The Stockholm+50 meeting was held in Stockholm, Sweden, on 2 and 3 June 2022, with the theme: "Stockholm+50: a healthy planet for the prosperity of all – our responsibility, our opportunity". The summit was held 50 years after the first Earth Summit in Stockholm in 1972. UN Secretary-General António Guterres summed up the entire situation in a few words when he requested all stakeholders to "end our suicidal war against nature".

The summit discussed the progress since the first Earth Summit in Stockholm in 1972, the need for political will to honour the existing commitments of countries, and the way forward. The members also agreed that financial and technological assistance should be provided to the developing countries if they are to meaningfully participate in their efforts towards sustainable development. It is hoped that this meeting of world leaders will help to renew the climate change commitments and result in instilling a new vigour to bring forth tangible action to prevent climate change and bring about sustainable development.

The Stockholm+50 meeting resulted in National Consultations facilitated by the UNDP. The *Global Synthesis Report of the National Consultations*[41] gives a clear idea of the efforts to achieve the UN sustainable development goals.

The Conference of Parties (CoP) is the decision-making body for any international conventions or treaties, with all member states being parties to it. COP for climate change under the UN has been in existence since 1995 when the first one was held in Berlin, Germany. The COP meets every year for their annual conference. Many decisions regarding climate change have been taken during these conferences. During every COP, the rich nations commit huge amounts of money for climate change actions. This includes advanced technologies, supporting less-developed nations, knowledge transfer, and even cash infusion. Unfortunately, these commitments are not met in full due to national priorities.

Recently, COP 28 was held at Dubai, UAE, from 30 November until 12 December 2023. The most important commitment at the COP 28 was a common agreement for transition away from fossil fuels. As per reports, more than $83 billion has been mobilised. It is to be hoped that in the coming years, the COP is able to make a more meaningful contribution to the global efforts to combat climate change.

The Conference of Parties[42] meets annually to measure the progress in tackling climate change and decide on the response so that global warming is prevented by stabilising GHG emission.

Now that we're aware of the history of global agreements to address climate change, let's look at the progress worldwide to date.

9.2 PROGRESS IN CLIMATE ACTION

The Climate Action Tracker (CAT) is an independent scientific project that tracks climate action taken by the Administrations and measures these actions against the Paris Agreement aims.

As can be observed in the projections shown in Figure 9.1, with the current policies and actions, global warming is expected to be between 2.5 and 2.7°C above the preindustrial limits of 1850. The 2030 target will limit this to 2.5°C. If the binding long-term targets and pledges are adhered to, global warming would be limited to 2.1°C. An optimistic scenario would be an increase of 1.8°C above the preindustrial limits.

Climate Action Tracker analyses data from across the world to arrive at a realistic assumption of the global warming projection. They estimated that when countries and industries achieve binding and long-term net-zero targets, the global warming would be limited to 2.1° above preindustrial limits.

Further to the Paris Agreement, there has been a noticeable action to tackle the GHG emissions. But it is still below the ambitions set and targets to be achieved. There was a sharp drop in GHG emissions during the coronavirus pandemic, but did not make any difference to the rising trend of emissions globally. There is no doubt that global warming and GHG emissions are firmly on the radar. Countries have legislated actions, and corporates are doing their bit towards achieving this goal. Fossil fuels are slowly coming into the limelight as polluters and alternative solutions are fast coming up. Technological advances have made renewable energy such as solar power and wind power affordable.

The graph in Figure 9.2 shows growth in the use of wind and solar power as renewable sources of energy from 2017 to 2021, as per the International Renewable Energy Agency (IRENA). It can be seen that the use of wind

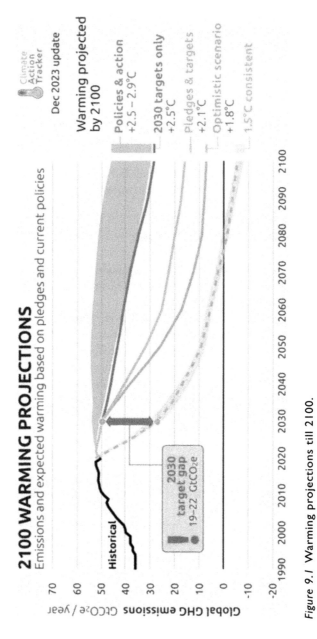

Figure 9.1 Warming projections till 2100.

Source: Climate Action Tracker[43]

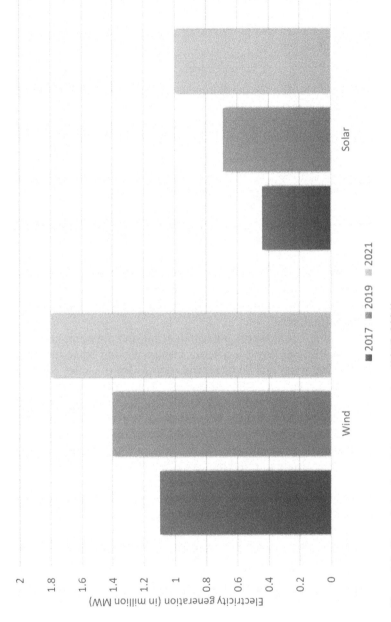

Figure 9.2 Growth in electricity generation (2017–2021).

Source: Irena.org[44]

power and solar power has grown significantly over the last few years. This is good news, but more needs to be done as the use of renewable energy is still relatively minimal.

IRENA is an intergovernmental agency that deals with energy issues. They support countries in their energy transition to cleaner forms of energy and promote the use of renewable energy. It has 168 member states and the European Union as members.

Now let us have a look at the development goals specified by the United Nations in their endeavour towards sustainable development. This is important as in their quest for economic growth and prosperity, other aspects such as health and the environment are often overlooked.

9.3 UN SUSTAINABLE DEVELOPMENT GOALS

In 2015, the United Nations Sustainable Development Summit established the United Nations SDGs[45] as an integral component of the 2030 Agenda for Sustainable Development. These consist of 17 goals, accompanied by 169 targets, which were established to succeed and advance upon the Millennium Development Goals (MDGs) endorsed by governments in 2001. These goals primarily focus on health, education, gender equality, and climate action, among others.

These SDGs address various environmental, social, and economic challenges facing the world and aim towards a more sustainable and equitable future. These goals include poverty and hunger eradication, access to quality education and healthcare, gender equality, combating climate change, protecting biodiversity and ecosystems, and promoting sustainable living. Each goal is accompanied by specific targets and indicators to track progress over the next decade.

SDG 7, which focuses on "Affordable and Clean Energy", is particularly important for the shipping industry. This SDG aims to "ensure access to affordable, reliable, sustainable and modern energy for all". For the shipping industry, this indicates a transition to cleaner fuels, improving energy efficiency, the use of renewable energy as a power source, continuing research and development of advanced technologies to reduce emissions, and, last but not least, regulatory compliance.

In order to encourage initiatives to reduce GHG emissions, the UN set up the GHG Compensation Fund to finance projects working towards GHG reduction.

9.4 GHG COMPENSATION FUND

The GHG Compensation Fund is a financial mechanism established to address the impact of GHG emissions on the environment. This fund serves as a means of mitigating the negative effects of carbon emissions and other GHGs that contribute to climate change. Contributions to the fund are typically made by industries, organisations, or individuals responsible for emitting these gases.

Established in 2010 by the United Nations Climate Change Conference, the primary purpose of the GHG Compensation Fund is to invest in projects and initiatives that promote environmental sustainability, carbon offsetting, and emissions reduction. These projects can include reforestation efforts, renewable energy installations, energy efficiency improvements, and other activities aimed at offsetting or reducing the overall carbon footprint. The fund operates by collecting monetary contributions from entities that emit GHGs beyond certain regulatory limits. These contributions are then allocated towards approved projects that have been vetted for their positive environmental impact. By channelling financial resources into such projects, the GHG Compensation Fund plays a role in encouraging the transition to a low-carbon economy and supporting global efforts to combat climate change. Overall, the GHG Compensation Fund acts as a mechanism for those who emit GHGs to take responsibility for their carbon footprint and actively participate in initiatives that counteract the negative effects of their emissions on the planet.

Ships are essentially GHG emitters and thus do not receive compensation from the GHG Compensation Fund. However, the shipping industry can gain indirect benefits from participating in the GHG Compensation Fund. They can finance projects that reduce emissions in other sectors, participate in carbon credit programmes, and support policies related to emission reduction.

As a regulatory measure in the maritime sector, EEDI and EEXI are perhaps the most important in our march towards zero-carbon shipping by 2050. Let us have a look at these requirements put forth by the International Maritime Organization (IMO) with the common aim of reducing GHG emissions.

9.5 IMO LEGISLATION

IMO has been consistently at the forefront of the UN's efforts to curb GHG emissions in the shipping industry. IMO regulations primarily focus on maritime safety and environmental protection. They also play a significant role in mitigating climate change by reducing GHG emissions from ships and promoting energy efficiency in the maritime sector. They have an important role in promoting energy efficiency in the maritime sector by setting standards, encouraging innovation, and promoting best practices in ship design, operation, and management. By driving improvements in innovative technologies to reduce fuel consumption and encouraging the transition to alternative fuels and renewable sources of energy, these regulations contribute to a more sustainable and environmentally friendly shipping industry. To spearhead these efforts, IMO has brought out several MEPC resolutions, each having the common goal of reducing GHG emissions. MEPC resolutions are decisions or recommendations adopted by the MEPC, a subsidiary body of the IMO. MEPC is tasked with the responsibility for addressing marine environmental issues. These resolutions provide guidelines, measures, strategies, standards, and procedures related to the prevention and control of pollution from ships and the promotion of sustainable maritime practices. Let us have a look at these initiatives.

9.5.1 Resolution MEPC 203(62) and Related Resolutions

Resolution MEPC 203(62)[46] was one of the first official documents to highlight energy efficiency as a mandatory requirement. Adopted on 15 July 2011, the resolution introduced amendments to the MARPOL Annex VI with the inclusion of a new chapter 4 (Regulations on Energy Efficiency for Ships).

MEPC 203 (62) contains the following:

- Requirements for International Energy Efficiency Certificate (IEEC):
- Regulations for attained EEDI and Required EEDI (for details on this topic, refer to Chapter 3). The attained EEDI should be less than or equal to the required EEDI. The specifics of this are given in the resolution.
- Ship Energy Efficiency Management Plan (SEEMP)

This resolution details the required EEDI, including the reduction factors for EEDI relative to the reference line.

IMO has issued several guidelines for the calculation of attained EEDI, keeping in mind the later requirements and the advances in technology, as follows:

- **MEPC.308(73)**: Guidelines on the Method of Calculation of the Attained EEDI for New Ships. Adopted on 26 October 2018, this resolution provides guidance on how to calculate the EEDI for new ships, ship designers, builders, and regulators can effectively evaluate the energy efficiency of new vessels by following these guidelines and adopt innovative technologies and practices to reduce emissions and enhance sustainability in the maritime industry.
- **MEPC.332(76)**: Amendments to the 2018 Guidelines on the Method of Calculation of the Attained EEDI for New Ships (Resolution MEPC.308(73), as Amended by Resolution MEPC.322(74)). Adopted on 17 June 2021, this resolution further clarifies the calculation process of the attained EEDI for new ships with the inclusion of a standard

format to submit EEDI information to be included in the EEDI database maintained by the IMO.

- **MEPC.364(79):** 2022 Guidelines on the Method of Calculation of the
  Attained EEDI for New Ships.[47] This resolution took into consideration the above guidelines with the various amendments and consolidated them along with new inputs. The resolution was published on
  1 January 2023, and became applicable to new ships thereafter, and
  superseding the earlier resolutions on the calculation of EEDI.

9.5.2 Resolution MEPC 278(70)

As we have seen, the entire concept of reducing GHGs revolves around fuel
consumption. Thus, it is important to know the fuel consumption of each
ship in order to calculate their CO_2 emissions. Consequently, the IMO introduced a DCS to monitor the fuel consumption of the ships.

This resolution was adopted in October 2016 and required ships to record
and report their fuel consumption data from 1 January 2019 (ships of 5,000
GT and above). The data is reported to the flag state or their RO at the end
of each calendar year. The flag state issues a Statement of Compliance to the
ship by 31 May of the next year. This Statement of Compliance is valid for
one year, that is, till 31 May of the following year. The flag state collects and
collates the data from the various sources and transfers the data to the IMO
Ship Fuel Consumption Database.

9.5.3 Resolution MEPC 80

IMO Resolution MEPC 80[48] was convened between 3 and 7 July 2023 and
adopted the 2023 IMO strategy on the reduction of GHG emissions from
ships. It can be considered to be the game-changer in the IMO's efforts
to combat GHG emissions and reduce global warming and paves the way
forward for MEPC Resolutions 81 and 82. It also formed the intersessional
working group (ISWG) on the Reduction of GHG Emissions from Ships
(ISWG-GHG 16). This will be discussed later in this chapter.

The 2023 IMO GHG Strategy is considered a significant step forward and
is deemed sufficiently aligned with the Paris Agreement's 1.5°C target. The
strategy aims to reduce GHG emissions by 20% striving for 30% by 2030,
70% striving for 80% by 2040, and reaching net-zero by 2050. It establishes a timeline for adopting measures and updating the strategy, with a
review scheduled every five years, the next one due in 2028. Current efforts
focus on implementing the outlined commitments and initiatives, with a
priority on establishing a pricing mechanism for maritime GHG emissions
to finance decarbonisation efforts.

MEPC 80 puts into perspective the entire strategy of the IMO regarding
the reduction of GHG emissions from ships and is linked with MEPC 81,

which we will discuss later in this chapter. This resolution covers a wide range of critical issues concerning maritime environmental protection and sustainability:

- **Tackling Climate Change:** The MEPC 80 session adopted the 2023 IMO Strategy of reduction of GHG emissions by at least 20%, striving for 30% by 2030 as compared to 2008 levels, leading to an ultimate aim of net-zero emissions by 2050. It was also agreed that the 16th session of the Inter-Sessional Working Group on the Reduction of GHG Emissions from Ships (ISWG-GHG 16) will be scheduled to meet in 2024 before the MEPC 81 session. MEPC 80 also agreed to include the modalities of on-board carbon capture in the ISWG-GHG 16. In addition, it adopted the guidelines on the life cycle GHG intensity of marine fuels (LCA guidelines) leading to well-to-tank and tank-to-wake emission factors, known as well-to-wake emissions.
  The committee also approved interim guidelines on the use of biofuels.
- **Energy Efficiency:** MEPC 80 approved the draft amendments to the IMO data collection system (DCS). The aim is to improve monitoring and reporting of ship fuel oil usage in order to analyse the fuel consumption of the ships. The committee also conducted a review of the measures for the short-term GHG reduction, EEXI, CII, and enhanced SEEMP, which entered into force on 1 November 2022. These are clarified by IMO in their EEXI and CII – ship carbon intensity and rating system.[49]
- **Reduction of Volatile Organic Compounds:** VOCs are released from fossil fuels carried as cargo and are considered GHGs. The resolution added a scope of work on the reduction of VOCs in order to reduce GHG emissions from tankers.

MEPC 80 also covers other topics such as ballast water management, biofouling management, designation of Particularly Sensitive Sea Areas (PSSA), underwater noise, tackling marine litter, ship-to-ship transfer, designation of special areas, and other miscellaneous matters.

MEPC 80 decided to conduct a comprehensive impact assessment (CIA). This would help in understanding the effect on the member states and the shipping industry so that further action can be taken to reach the ambition of net-zero shipping in 2050 without adversely affecting the interests of the stakeholders involved. For facilitating the CIA, MEPC 80 invited the Secretary-General of the IMO to create a steering committee to study and prepare a CIA of the basket of mid-term measures on the states. The revised procedure for assessing these impacts is detailed in MEPC.1/Circ.885/Rev.1.[50]

This MEPC circular gives an understanding of the CIA and specifies the analysis tools, models, and support for the assessment. The CIA is scheduled to be tabled during the MEPC 81 to be held in March 2024.

9.5.4 Resolution MEPC 81

The 81st session of the MEPC was held between 18 March and 22 March 2023. The agenda included many aspects of environmental protection, but the focus was on the GHG fuel intensity requirements along with a GHG pricing mechanism. Other approvals were granted such as ECAs, amendments to the Ballast Water Convention, potential extension, revision of the NO$_x$ requirements, guidelines for sampling of fuel oil, and revision of procedure for exhaust gas cleaning systems.

The main focus of MEPC 81[51] was energy efficiency, and the following were some of its outcomes:

- During the 80th session of the Marine Environment Protection Committee (MEPC 80), proposed changes to Appendix IX of MARPOL Annex VI were endorsed, updating and expanding the data required for reporting within the IMO DCS. At the MEPC 81, these proposed amendments from MEPC 80 were officially adopted, ushering in additional reporting requirements such as reporting total fuel oil consumption per combustion system and total fuel oil usage during periods when the vessel is idle. In addition, fuel to be reported per combustion system (main engines, auxiliary engines, and so on), total amount of shore power supplied, energy efficiency technologies utilised, and so on.
- Revisions to the pertinent IMO SEEMP guidelines were ratified. These amendments encompass specifications for determining "total transport work", which is computed according to the Guidelines concerning the Shaft/Engine Power Limitation System to Comply with EEXI Requirements and the Use of a Power Reserve[52].
- During this session, in order to comply with the EEXI regulations commencing in 2023, they were reviewed. Consequently, guidelines were issued allowing the unlimiting of the Shaft Power Limitation (ShaPoLi) or Engine Power Limitation (EPL) system under specific circumstances endangering the vessel's safe navigation. Additionally, new criteria were incorporated for ShaPoLi systems, ensuring their control remains independent of engine automation.
- On-board Carbon Capture Systems (OCCS) efforts were initiated to advance on-board carbon capture (OCC) technologies aimed at mitigating GHG emissions by isolating and capturing CO_2 from exhaust gases on-board vessels. During the previous session, a new agenda item was approved within the ISWG on reduction of GHG from ships (ISWG-GHG) to formulate a regulatory framework for this purpose. In the current session, it was decided to draft a comprehensive work plan for the development of regulations pertaining to OCC technologies within the newly established Correspondence Group.

MEPC also discussed other matters and issued relevant instructions and guidelines. These include temporary storage of treated sewage and grey water, designation of NO_x and SOx ECAs, review of BWM convention, and other miscellaneous matters.

9.5.5 The Intersessional Working Group on Reduction of GHG Emissions from Ships (ISWG-GHG 16)

The ISWG on reduction of GHG emissions from ships, a specialised body under the IMO, focuses on developing measures to reduce emissions from the shipping industry. Meeting between sessions of the MEPC, it plays a vital role in advancing discussions and proposing regulatory frameworks to address this pressing issue. Through collaboration and technical expertise, the group works to promote cleaner technologies, improve operational efficiency, and encourage the use of alternative fuels within the maritime sector, aligning with broader international efforts to mitigate climate change and achieve sustainability goals.

The 16th meeting of the ISWG-GHG 16 was held between 11 and 15 March 2024 and the report was presented to MEPC 81 on 18 March 2024.[53] The Group pledged to develop a GHG fuel standard to regulate the reduction of GHG emissions as well as a pricing mechanism to incentivise measures to reduce these emissions. The Steering Committee established by the Secretary-General of IMO presented the report on the CIA of the basket of mid-term measures. Based on this, the Group decided to recommend a two-day expert workshop (GHG-EW 5) on further development of mid-term measures for GHG reduction. Their outcome is to be reported to MEPC 82, to be held in the autumn of 2024.

The working group also finalised the draft 2024 Guidelines on Life Cycle GHG intensity of marine fuels (LCA Guidelines) to facilitate well-to-wake calculation of GHG emissions. The proposal related to on-board capture of CO_2 was considered, including the development of a regulatory framework for the same.

In conclusion, we can say that global policymakers are becoming aware of the environmental damages caused over the last few decades in the name of development. The catchphrase now is sustainable development. It is to be seen whether these conferences and summits lead to some tangible policy decisions. Meanwhile, IMO has been marching ahead with its GHG reduction strategies. Since it has the power to make mandatory regulations and enforce them through the various port state controls and flag state controls, the shipping industry has no option but to fall in line.

In general, MEPC Resolutions play a crucial role in reducing emissions from ships by developing and implementing regulations, promoting energy efficiency measures, establishing ECAs, monitoring emissions, promoting alternative fuels and technologies, and advancing a global strategy to address climate change in the maritime sector.

9.6 SUMMARY

In this chapter, we have discussed the various international legislations in force and how they contribute to the efforts of the global community to reduce global warming. The United Nations have organised several summits and conferences in order to make the member countries aware of the dangers associated with the emission of GHG and the ways and means to control the same. This chapter sheds light on the Earth Summits, the Kyoto Protocol, the Paris Agreement, as well as the recent COP. The GHG fund set up specifically to address the impact of GHG emissions is also discussed. The relevant MEPC resolutions put forth by the IMO have been discussed in order to fully understand their implications.

In the next chapter, we will be discussing the efforts by the shipping industry towards reducing the GHG emissions and make shipping greener and more sustainable. Ship-owners and other stakeholders have been working together for a long time to achieve this common goal; these initiatives will be highlighted and discussed.

BIBLIOGRAPHY

36. American Institute of Physics. *Timeline of Global Warming*. https://history.aip.org/climate/timeline.htm.
    The entire timeline of global warming up to 2024 is given here.
37. Stockholm Declaration Is a Proclamation by the United Nations at the Stockholm Meet from 5 to 16 June 1972. https://wedocs.unep.org/bitstream/handle/20.500.11822/29567/ELGP1StockD.pdf.
    This declaration underlines the commitment of the Unites Nations to uphold sustainability and environmental protection along with economic necessities.
38. Agenda 21. (1992). https://sdgs.un.org/sites/default/files/publications/Agenda21.pdf.
    Agenda 21 was released during the Earth summit 1992. It is a comprehensive plan of action to be taken for sustainable development.
39. Rio + 20: The Future We Want. (2012). https://sustainabledevelopment.un.org/futurewewant.html.
    This document released on the occasion of the United Nations Conference on Sustainable Development known as Rio+20 is a declaration by the attending world leaders affirms their vision towards sustainable development and the roadmap to achieve the same.
40. UNFCC Report on the Paris Agreement. (2015). https://unfccc.int/sites/default/files/english_paris_agreement.pdf.
    This report details the Paris Agreement between the parties to the UNFCC convention in order to foster engagement between the parties and achieve the UN sustainable development goals.
41. UNDP Report on Stockholm +50 National Consultations. (2022). https://www.undp.org/sites/g/files/zskgke326/files/2022-11/UNDP-Stockholm-50-A-Global-Synthesis-Report-of-National-Consultations_0.pdf.

This Global Synthesis Report on the Stockholm+50 National Consultation was facilitated by the UNDP involving 56 developing countries.

42. Conference of Parties. (2023). https://www.un.org/en/climatechange/un-climate-conferences.

The conference of Parties is the response of all parties in order to take concerted actions to tackle climate change.

43. Climate Action Tracker. (December 2023). *2100 Warming Projections: Emissions and Expected Warming Based on Pledges and Current Policies.* https://climateactiontracker.org/global/temperatures/. Copyright ©2023 by Climate Analytics and New Climate Institute. All rights reserved.

The Climate Action Tracker gives the projections for global temperatures by estimating emissions and expected global warming based on current policies and pledges by the stakeholders.

44. Transition to Wind Energy. (2023). https://www.irena.org/Energy-Transition/Technology/Wind-energy.

Contains data and statistics by IRENA related to the use of wind energy as a source of power. Also contains the average costs for commissioning of wind projects.

45. The Seventeen Sustainable Development Goals of United Nations. (2023). https://sdgs.un.org/goals.

These are the United Nations goals to eradicate poverty, protect the planet and ensure that there is peace and prosperity all around.

46. IMO Resolution MEPC203(62). (Adopted on 15 July 2011). *Energy Efficiency Regulations Included in MARPOL Annex VI.* https://wwwcdn.imo.org/local resources/en/OurWork/Environment/Documents/Technical%20and%20Oper ational%20Measures/Resolution%20MEPC.203%2862%29.pdf.

This IMO resolution introduced energy efficiency and the SEEMP.

47. IMO Resolution MEPC.364(79). (1 January 2023). *2022 Guidelines on the Method of Calculation of the Attained Energy Efficiency Design Index (EEDI) for New Ships.* https://wwwcdn.imo.org/localresources/en/KnowledgeCentre/IndexofIMOResolutions/MEPCDocuments/MEPC.364(79).pdf.

This IMO resolution gives the formula for calculation of attained EEDI for new ships. Correction factors are included based on ship types.

48. IMO Resolution MEPC 80. *2023 IMO Strategy on Reduction of GHG Emissions from Ships.* https://www.imo.org/en/MediaCentre/MeetingSummaries/Pages/MEPC-80.aspx.

This IMO resolution details the 2023 strategy of IMO for reducing emission of GHGs from ships.

49. EEXI and CII – Ship Carbon Intensity and Rating System. (2019). https://www.imo.org/en/MediaCentre/HotTopics/Pages/EEXI-CII-FAQ.aspx.

The requirements of the annual CII are explained in detail.

50. Revised Procedure for Assessing Impacts on States of Candidate Measures. (2023). https://wwwcdn.imo.org/localresources/en/OurWork/Environment/Documents/MEPC.1-Circ.885-Rev.1.pdf.

This IMO circular describes the procedure for conducting the comprehensive impact assessment by the steering committee. It also explains the establishment, role and function of the steering committee.

51. Report of the MEPC 81. (2024). https://www.imo.org/en/MediaCentre/MeetingSummaries/Pages/MEPC-81.aspx.

This report details the results of the 81st session of IMO. MEPC 81 is seen as taking forward the requirements of MEPC 80.

52. IMO Resolution 335(76) (Adopted on 17 June 2021). *Guidelines Concerning the Shaft/Engine Power Limitation System to Comply with EEXI Requirements and the Use of a Power Reserve Actual.* https://wwwcdn.imo.org/localresources/en/OurWork/Environment/Documents/Air%20pollution/MEPC.335(76).pdf.

This resolution details the general system requirements for ShaPoli/EPL as well as the use of power reserve for unlimiting the engine power limitation.

53. Report on the ISWG on reduction of GHG emissions from ships (ISWG-GHG 16). (11–15 March 2024). https://www.imo.org/en/MediaCentre/MeetingSummaries/Pages/Intersessional-Working-Group-on-Reduction-of-GHG-Emissions-from-Ships-(ISWG-GHG-16),-11-15-March-2024.aspx.

Chapter 10

Environmental Efforts by the Shipping Industry

The shipping industry has been taking various voluntary measures to contribute to environmental efforts and address sustainability concerns. There is no doubt that it has to comply with mandatory requirements. However, many shipping companies have openly supported the drive towards reducing greenhouse gas (GHG) emissions and initiated various awareness programmes to sensitise their ships' crew towards energy efficiency.

This chapter discusses International Maritime Organizations (IMO's) initiatives to support developing countries, least developed countries (LDCs), and small island developing states (SIDS) in their endeavour to reduce GHG emissions in order to restrict global warming. It is important to know about these initiatives in order to understand the steps taken by IMO towards sustainable development and their impact on the shipping industry. The efforts of the ship-owners in ordering innovative ships that depend on alternate fuels and renewable energy, thereby reducing their footprint, are also discussed.

Developed countries are on relatively better ground where the reduction of GHG emissions is concerned. This is because technological advances are more cost-effective in developed countries than in developing countries. This is the reason such countries require hand-holding in the form of financial incentives, technology transfers, and so on. Let us discuss the initiatives by IMO to support these countries.

10.1 IMO'S SUPPORT FOR DEVELOPING STATES AND LDCs

IMO has a comprehensive programme of support for developing States to implement IMO regulations. These countries require support for capacity building and technical assistance in order to comply with the regulatory requirements. Further, they may face challenges in implementing emission reduction methods, reducing environmental pollution, and so on. With their various initiatives, IMO aims to support these countries in building sustainable maritime sectors and thereby promote economic development

DOI: 10.1201/9781032702568-10

without degrading the environment or compromising the quality of life of their populace.

Some of the global projects specifically targeting GHG reduction measures include the following:

- **IMO's Integrated Technical Cooperation Programme (ITCP):** This programme was initiated by IMO to provide assistance to developing countries in building up their human and institutional capacities in order to comply with the regulatory requirements. Countries in Africa, SIDS and LDCs have been identified as those countries that most require such assistance. Towards this end, IMO secures financial, human, and logistical support. To further this, IMO's ITCP was formed to promote sustainable and socio-economic development in these countries.
- **IMO GHG Technical Cooperation Trust Fund:** MEPC 74 established this multi-donor trust fund in May 2019, in order to provide financial and capacity-building support for the implementation of the Initial IMO Strategy on GHG emissions. This fund is primarily aimed at developing countries in order to incentivise them to comply with the requirements. Essentially the assistance is to help the developing countries enhance their institutional and technical capacity and facilitate the transfer of technology related to the reduction of GHG emissions so that they can implement the necessary measures required by the IMO's strategy.
- **GreenVoyage2050[54]:** This is a partnership project between the IMO and the government of Norway to support developing and least developed countries to participate in the climate change initiatives of IMO. The project was launched in May 2019 with the common aim of reducing the carbon footprint of ships. The project recently got a fresh lease of life to continue its work through 2030. This was mainly due to an injection of USD 19.4 million by Norway for phase two of the project. The ambition of the project is to involve other developed countries to fund the initiative and support the aim of working towards decarbonisation of the global shipping.
- **Sustainable Maritime Transport Training Programme (SMART):[55]** This is an initiative by IMO to assist LDCs and SIDS by means of enhanced training programmes. The programme is funded by the Republic of Korea (RoK) and aims to increase the human capacity in various aspects of IMO's GHG Strategy. By means of comprehensive training, monitoring, and evaluation, the personnel involved in maritime transport are trained in global regulations including MARPOL, alternate fuels for the decarbonisation of shipping, best practices, and the use of technologies for complying with the IMO strategy for reduction of GHG emissions. The aim is to give training and support so that they can effectively implement the IMO GHG strategy.

Recently the SMART-C Framework Agreement was signed on 28 November 2023, where RoK agreed further funding of the programme.

- **Global Maritime Technologies Cooperation Centres (MTCC) Network:** IMO has been working round the clock to make all stakeholders participate in their GHG strategy towards reduction of GHGs and increase the energy efficiency of ships. Towards this end, they have connected with many member states, ports, ship-owners, and other interested parties. One such initiative is the Global MTCC Network, which, as the name suggests, is a global network for energy-efficient shipping. The project is funded by the European Union and sets up MTCC in various regions, especially LDCs and SIDS. These MTCCs are selected to become centres of excellence with the common aim of ensuring compliance with relevant energy efficiency regulations, help policy making in this regard, encourage use of low-carbon technology in maritime transport, and set up a robust feedback and monitoring system. To further their objective, MTCC connects with major academic institutions, government departments, shipping companies, international chartering companies, and so on.

  MTCC provides technical assistance, knowledge sharing, and efforts to promote the adoption and application of energy-efficient technologies and practices. The network contributes to reducing fuel consumption, GHG emissions, and environmental impact in the global shipping industry by promoting research, development, and international cooperation between stakeholders, fostering a more sustainable future for maritime transportation.

- **IMO Resolution MEPC 323(74):** In order to facilitate cooperation between stakeholders, IMO adopted Resolution MEPC 323(74)[56] on 17 May 2019. This resolution is an invitation to all member states to foster cooperation between port and shipping sectors with the objective of reducing GHG emissions from ships. Based on statistical analysis, IMO realised that forcing regulations on unwilling ship-owners would only be successful to a certain extent. Flag administrations, various ports, and other stakeholders, such as shipping organisations, classification societies, and so on, needed to be involved in order to work holistically and achieve the success of IMO's GHG strategy. This resolution encourages member states to promote incentive schemes aimed at the reduction of GHG emissions. Further cooperation between interested parties such as ports, shipping companies, and even bunker suppliers is encouraged as this is the only way to the net-zero ambition of IMO.

IMO has moved far ahead in encouraging cooperation between various parties and has initiated several partnership programmes (in addition to the ones we've just discussed) with countries and other stakeholders. These partnership and projects[57] are aimed at encouraging the involvement of all concerned with the objective of reducing GHG emissions.

Let us now have a look at how shipping companies and ships are rewarded for their energy efficiency efforts. Such recognition comes with both financial, commercial, and social benefits. Ships and companies that are conferred these awards are seen as contributing to sustainable shipping and therefore stand tall, apart from the others.

10.2 ENERGY CONSERVATION AWARDS

Ships navigating the oceans consume vast quantities of fuel, releasing significant amounts of CO_2 and other harmful gases into the air. Although advancements in technology offer the potential to decrease fuel consumption and emissions, shipping companies are hesitant to lead in these efforts primarily due to the substantial cost of such initiatives, coupled with minimal cost-benefit for ship-owners in complying beyond basic requirements. To encourage ship-owners to embrace energy conservation and reduce GHG emissions, the IMO, United Nations, and various other organisations have introduced the concept of awards to top performers in this realm. Some of these awards come with monetary incentives, while others provide recipients with a competitive edge in the industry. For instance, in numerous US ports, ships honoured with Green Awards receive discounts on port dues, and others enjoy advantages such as reduced waiting times for pilots. Each of these awards offers both tangible and intangible benefits, aiming to motivate ship-owners to adopt measures that mitigate environmental impact.

10.2.1 Green Award

This is a non-profit international organisation established in 1994 in Rotterdam. Its aim is to provide incentives for ships that conform to standards that are over and above the regulatory requirements. It is a voluntary quality assessment scheme. They inspect the ships for safety, quality, and environmental performance, and if the ship performs satisfactorily, a Green Award certificate is issued.[58] This certificate is evidence that the company is in compliance with the IMO requirements regarding GHG emissions, and these ships contribute to sustainable shipping. Green Award is quite popular in the industry, with many shipping companies, ports, and service providers opting to be part of the initiative for greener shipping. Green Award-certified ships are entitled to several incentives, such as discounts on port dues, products, and services, enhanced reputation, and preference by charterers and shippers. Let's look at these benefits in greater detail:

- **Cash Benefits:** Some of these awards carry a cash award for the ship and ship-owner. These awards may be offered by government agencies, industry associations, or non-profit organisations aiming to promote sustainability in the shipping industry. Reduction in port dues is also one of the benefits of such awards.

- **Canal Dues:** Major canal authorities, such as Suez Canal and Panama Canal, have introduced a carrot-and-stick policy for ships transiting their canals. Green ships are entitled to a reduction in dues, while high-emitting ships are penalised with higher fees.
- **Reduction in Insurance Premium:** Insurers who sign up to the Poseidon Principles have to disclose the climate alignment of their hull and machinery portfolios. In order to ensure that this meets the IMO targets, they offer a reduction in insurance premiums to those ships who are in compliance with the IMO requirements and are in possession of the Green Award certificate. The Poseidon Principles will be discussed in the next chapter.
- **Preference in Obtaining Lucrative Charters:** Tanker majors and other major charterers give preference to those ships who are holders of the Green Award certificate. This is an incentive for shipping companies to go that extra mile and ensure that their ships are top performers in energy efficiency. They can thus obtain lucrative charters and command better charter hire than other ships.
- **Increase in Protection and Indemnity (P & I) Club Benefits:** P & I clubs are also part of the overall global efforts to reduce global warming and GHG emissions. To this end, they offer benefits to those ships who have been awarded Green Award certificates. The benefits may include lower premiums, increased coverage, and so on.
- **Public Relations Benefits:** Obtaining a Green Award has several PR benefits. It is evidence of the company's commitment to environmental responsibility; it gives the ship a positive public image wherever she travels and enhances the stakeholder's confidence in the company.
- **Increase in the Morale of Ship Staff:** Employees take pride in being part of companies that prioritise sustainability. Obtaining a Green Award certificate will boost the morale of the ship staff and motivate them to work further towards ensuring energy efficiency and reducing GHG emissions.
- **Reduction in Pilotage Charges:** Many ports offer a discount on pilotage, mooring, and berth hire for ships with Green Award certificates. Since these charges are substantial, the discounts can be quite large and may benefit the bottom line of the ship-owner. A list of incentives offered by various ports is given later.

Green Awards for ships encourage and recognise energy efficiency and sustainability efforts within the maritime industry. By assessing various aspects of a ship's operations, such as fuel consumption and emissions reduction, these awards encourage shipping companies to invest in eco-friendly technologies and practices. Recipients benefit from reduced operating costs, improved reputation, and access to potential financial incentives. Overall,

Green Awards serve as catalysts for promoting energy efficiency and environmental responsibility in the maritime sector.

10.2.2 Clean Shipping Index

The clean shipping index (CSI) measures the environmental compliance of the ship. Ships that demonstrate their commitment to reducing emissions beyond regulatory requirements can earn the CSI certificate. This certificate is evidence of the commitment of the ship and the ship-owner towards sustainable practices over and above the IMO strategy. The benefits of the CSI are somewhat similar to those of the Green Awards and are as follows:

- Cost savings due to more fuel-efficient operations, streamlined processes, and so on
- Environmental impact reduction by reducing GHG emissions
- Regulatory compliance by encouraging ships to adopt practices that are in line with current regulations
- Market reputation is enhanced, as companies that achieve a high CSI score indicate a commitment to sustainable operations
- High CSI score also improves the motivation of the employees and ensures that they perform better.

The CSI facilitates energy efficiency in the shipping industry by promoting the adoption of cleaner technologies and practices, promoting investment in fuel-saving measures, and collaboration among stakeholders. Through its transparent evaluation framework, the CSI encourages shipping companies to improve their energy performance, driving innovation and progress towards a more sustainable maritime sector.

Classification societies[59] provide guidance and services to obtaining the CSI index. It is mainly voluntary, but some ports in Sweden, Finland, and Canada have applied the CSI in their operations.

10.2.3 Ports Environmental Review System

The ports environmental review system (PERS) certification is awarded to ports that demonstrate continuous improvement in environmental protection and energy efficiency measures. Ports that successfully implement and maintain environmental management systems are awarded these certificates. Ships calling at such ports will also have to adhere to higher environmental standards.

There are several benefits for ports that participate in the PERS programme and obtain a certification of verification:

- **Market Reputation:** Ports that obtain the certificate will enjoy a better reputation as a port that is committed to maintaining environmental standards.

- **Competitive Advantage:** In regions like Europe with strict environmental regulations or where the customers value environmental sustainability, ports that have been awarded the PERS certificate have an advantage over other ports in attracting ship-owners and their ships.
- **Funding and Incentives:** Ports awarded with the PERS certificate of verification will be more likely to obtain grants, subsidies, and other financial incentives.

These are some of the obvious benefits to ports, but there are other intangible benefits that accrue to companies that maintain environmental, social, and governance (ESG) standards. These standards will be discussed in Chapter 14.

The PERS contributes to energy efficiency in ports by providing a methodology for evaluating and enhancing environmental performance. Through assessment of energy consumption and emissions, PERS helps identify areas for improvement, guiding ports towards more sustainable practices. By encouraging collaboration and sharing best practices, PERS facilitates the implementation of measures to optimise energy usage and reduce environmental impact, ultimately promoting greater energy efficiency within port operations.

The adage "Charity begins at home" suggests that initiatives to decrease GHG emissions from ships will only be effective if ship-owners actively support and genuinely comply with these endeavours. Let us have a look at how some of the major shipping companies are moving ahead with their efforts to meet the IMO GHG strategy of zero-carbon ships well ahead of 2050.

10.3 SHIPPING COMPANIES IMPLEMENTING SUSTAINABLE PRACTICES

At the centre of the efforts towards decarbonisation of the maritime industry are the ship-owners and managers. The environmental footprint of shipping operations holds significant importance for shipping companies due to compliance obligations, their ethical commitment to ESG principles, and the improvement of their commercial viability. Hence, they should, as a matter of policy, allocate a portion of their earnings towards research and development. This investment aims to discover new methods that allow for sustainable shipping practices, minimising environmental harm. Some of the shipping companies that are making efforts to reduce the footprint of their ships are:

- **Nippon Yusen Kabushiki Kaisha:** Japan's foremost shipping company, Nippon Yusen Kabushiki Kaisha (NYK), is making serious efforts to reduce emissions in its transportation and logistics activities. These include the following:
  - In a pioneering move in 2008, NYK, in partnership with Nippon Oil, unveiled the world's first solar-assisted cargo carrier, named

the Auriga Leader. This vessel utilises solar-generated electricity to transport as many as 6,200 vehicles for the Toyota Motor Corporation. The on-board solar power system consists of 328 solar panels, capable of generating 40 kilowatts of power.

- NYK announced the "NYK Group Decarbonization Story" on 6 November 2023, specifying their decarbonisation strategies and GHG reduction targets towards 2050.
- The world's first LNG-fuelled coal carrier, the 95,223 tons deadweight Panamax vessel, was delivered on 2 October 2023. Built by NYK and Kyushu Electric Power company in Oshima shipyard, Japan, the vessel will have zero emission of SOx, an 80% reduction in emission of NO_x, and a 30% reduction in CO_2 emissions. More such vessels are due in the pipeline.
- **AP Moller-Maersk**: The Danish conglomerate, known in shipping circles as Maersk Line, is actively pursuing energy and carbon efficiency enhancements through various means. They have established concrete short-term objectives for 2030 to achieve substantial advancements in mitigating direct emissions from their operations within this decade. These objectives encompass a 50% decrease in emissions per transported container in the Maersk Ocean fleet and a 70% reduction in absolute emissions from fully controlled terminals. Depending on the growth in the ocean business, these targets will result in absolute reductions in emissions ranging from 35% to 50% compared to the baseline year of 2020. They aim to achieve net-zero-carbon emissions by 2040, a decade before the target set by the IMO GHG strategy. They have 24 container vessels on order, equipped with dual-fuel engines, capable of operating on green methanol. These vessels range in size from 9,000 to 17,000 TEU (20-foot equivalent units). In fact, from 2021 onwards, Maersk Line has implemented a policy of ordering new vessels that operate on green fuels.

  In addition to addressing its own direct emissions stemming from operations, the group is dedicated to aiding its customers in reducing carbon emissions within their transport and logistics supply chains.
- **Hapag Lloyd**: Hapag-Lloyd has reached a noteworthy achievement as the first shipping line to integrate electronic fuel injection and valve control systems into the main engines of its container vessels. This pioneering approach ensures cleaner fuel combustion, leading to a significant decrease in NO_x exhaust emissions. This technological advancement not only improves overall efficiency but also lowers fuel consumption, making a substantial contribution to the reduction of total emissions. The Hapag-Lloyd management aims to convert their entire fleet to carbon-zero by 2045, that is, five years ahead of the IMO timeline. As an interim target, they aim to reduce the carbon intensity of their fleet by 30% compared to 2019 levels by 2030.

According to Lutz-Michael Dyck, Senior Director Strategic Asset Projects,

[T]he really important decision is what the future fuels will be. A lot of things are possible, but it will cost a lot of time and money to get potential solutions to the point that they're ready to be used on an industrial scale. At the moment, ammonia isn't an option in container and passenger shipping, as the safety concerns are still simply too great. Methanol is the best option, but since it costs nearly three times the cost of fossil fuel with low sulphur content it will be difficult for us to shoulder the cost for climate-neutral propulsion on our own.

This statement was made while explaining Ship Green, an innovative concept to involve shippers in their climate-friendly shipment initiative; it clearly shows the dilemma of ship-owners in switching to carbon-neutral fuels. It also highlights the necessity of the entire industry to support such programmes so that the IMO GHG strategy can be achieved.

- **Mitsui O.S.K. Lines:** Mitsui O.S.K. Lines (MOL) has been at the forefront of reducing the carbon footprint of shipping and has implemented several innovative technologies and design approaches to improve the energy efficiency and environmental performance of their ships:
  - **Propeller Boss Cap Fins System:** The Propeller Boss Cap Fins (PBCF) system, a technology jointly developed by MOL in partnership with West Japan Fluid Engineering Laboratory Co. Ltd. and Mikado Propeller Co. Ltd., is engineered to harness energy from the flow-out energy within the propeller hub vortex of a ship's propeller. As a ship's propeller rotates, it generates a vortex in its wake, resulting in energy dissipation. The PBCF system tackles this challenge and provides numerous advantages including fuel savings, speed improvement, vibration and noise reduction, as well as balanced propulsion.
  - **Wind Challenger:** MOL announced that the wind-powered M. V. Shofu Maru was delivered and started operations on 2 October 2022. Built at the Oshima shipbuilding yard, the 100,422 tons deadweight coal carrier is equipped with the futuristic wind challenger for harnessing wind energy. The wind challenger uses a telescoping hard sail. The sail is fitted at the bow of the vessel and is robust enough to operate in severe weather conditions. The Shofu Maru is a hybrid vessel, which can also operate on normal fuel during unfavourable weather conditions and low wind.

    MOL's dedication to incorporating these technologies and design enhancements on their ships underscores their commitment to energy efficiency and environmental responsibility. These endeavours align with worldwide initiatives aimed at reducing GHG in maritime transport.
- **Bernhard Schulte Shipmanagement:** Bernhard Schulte Shipmanagement (BSM) will be rolling out an innovative IT platform designed to

streamline and manage the comprehensive compliance process associated with the impending introduction of the EU Emissions Trading System (ETS) in 2024. The EU ETS is the world's first and largest carbon trading market and is a complex system which requires expertise to navigate successfully. The platform will address key components of the compliance process, encompassing data collection, verification, emissions forecasting, registration management, and carbon allowance processes. BSM has outlined that the existing digital ship management system will be enhanced with a consolidated dashboard, providing a unified source for seamless oversight and control.

Shipping companies are increasingly utilising innovative technologies to enhance energy efficiency. These include dual-fuel engines, use of renewable sources of energy such as wind and solar power, and on-board energy management systems. Additionally, optimised hull designs and investments in alternative fuels like biofuels and hydrogen are further driving sustainability in the maritime industry. In addition, ship operators are increasingly involving ship staff in these measures through regular training in energy efficiency and regulatory compliance. By means of these initiatives, shipping companies are constantly endeavouring to reduce fuel consumption, lower emissions, and advance towards a more environmentally sustainable future.

Stakeholders in the shipping industry have a major role to play in ensuring that the decarbonisation efforts are successful. Let us examine the approaches and considerations driving the transition towards a more sustainable maritime sector and the role of their stakeholders in this transition towards a more sustainable future.

10.4 STAKEHOLDERS EFFORTS TOWARDS DECARBONISATION

Stakeholders' contributions are integral to achieve the goal of decarbonisation in the maritime industry. These contributions stem from compliance requirements, to ethical imperatives, and commercial viability. Let's look at some key stakeholders that play in these efforts:

- **Flag States:** The flag states have had a major role to play in safe and environment-friendly shipping. It is the responsibility of the flag states to ensure that the ships that are registered under their flag follow international bodies' guidelines and conventions. Shipping companies report their ship's fuel consumption, GHG emissions, and carbon intensity (CII) data to the flag administration, which is responsible for collecting, collating, and monitoring this information for all the ships flying their flag. This information has to be forwarded to the IMO for further analysis and inclusion in their database. The flag states are responsible for issuing the International Air Pollution Certificate (IAPP) and the

International Engine Air Pollution Certificate after the relevant initial or renewal surveys have been conducted, and the ship is found in compliance with the requirements of Annex VI of MARPOL 73/78. They also issue an International Energy Efficiency Certificate at the first IAPP intermediate or renewal survey after 1 January 2023. The flag state also confirms that the ship is complying with the EEXI requirements during the first IAPP intermediate or renewal survey after 1 January 2023.

By 31 March 2023, flag administrations will issue a Statement of Compliance to those ships that have been verified for satisfactory compliance with the attained CII. It is important to note that the RO may be permitted to issue statutory certificates on behalf of the flag administration.

- **Port State Control:** Port state control (PSC) involves a nation asserting control and authority over foreign ships within waters under its jurisdiction. A country has the right to establish its own regulations, imposing specific requirements on foreign vessels engaged in trade within its waters. Nations participating in international conventions are granted the authority to inspect ships from other countries operating in their waters to ensure compliance with the obligations outlined in these conventions, particularly those related to emissions and pollution. In fact, PSC officers have been known to board ships and confirm compliance with mandatory requirements such as Ship Energy Efficiency Management Plan (SEEMP), CII, and so on.

- **Classification Societies:** Classification societies play a major role in ensuring compliance with regulatory requirements. From the shipbuilding stage, classification societies are involved with the design and implementation of various standards, including ships' design features, equipment, and other systems to ensure they meet the efficiency criteria. Classification societies also work as RO authorised by the flag administration to inspect, survey, and issue statutory certificates on their behalf. Thus, they have an important role to play in the IMO 2023 strategy. They work in coordination with shipyards, ship-owners, flag state authorities, PSC, and other national and international bodies. Classification societies help ensure that ships comply with international and regional regulations related to energy efficiency, such as the IMO, EEXI, and the EEOI. They develop and update technical standards and guidelines that address energy efficiency measures for ships. They also track and report data related to fuel consumption, emissions, and other parameters, helping ship-owners and operators optimise their vessel's performance.

In an era of evolving regulations, ship-owners and operators are often confused about compliance procedures. In such a scenario, classification societies offer advisory services to ship-owners and operators on how to enhance energy efficiency and comply with regulations. This can include recommendations for retrofitting existing vessels with energy-efficient technologies, documentary requirements, and so on.

- **Ports:** Ports play a major role in assisting ships to increase their energy efficiency. They are involved in providing various services to the ships in this regard. Providing shore power to the ships once they are berthed will enable the ships to shut down their generators. This is known as cold ironing and will reduce air pollution and fuel consumption. Today several ships are operating on hybrid mode with the option of switching over to the alternate mode if required. If the port is not able to supply the required fuels, then the ships have to continue operating on HFO resulting in higher fuel consumption.

  Many ports provide incentives in the form of reduced port dues or lesser waiting periods for ships that adopt energy-efficient practices. In fact, some ports offer berthing on arrival to ships that operate on renewable energy sources like wind energy. In addition, having efficient waste collection and sludge collection systems will go a long way to protect the environment. Moreover, many ports themselves adopt energy-efficient practices in their day-to-day operations, thereby resulting in sustainable practices.

- **Shipbuilders and Designers:** Shipbuilders and designers play a vital role in energy efficiency. Ship operations are limited in their capacity to reduce fuel consumption. It is ultimately the shipbuilders and designers who can come up with innovative technical solutions to increase the energy efficiency of ships. This includes new types of hull design, propulsion systems, engine management systems, and so on, all of which contribute to increased energy efficiency. Designers can even come up with new technological solutions that can be retrofitted on existing ships in order to increase energy efficiency and reduce fuel consumption.

Stakeholders in the shipping industry, including ship-owners, shipping companies, shipbuilders, maritime regulators, port authorities, and international organisations like the IMO, collaborate to improve energy efficiency. Ship-owners and companies invest in fuel-efficient technologies, while shipbuilders develop innovative designs to make the ships more fuel-efficient. Regulators enforce standards, port authorities incentivise efficiency, and international organisations provide guidance. This collective effort drives progress towards a more sustainable and energy-efficient maritime sector.

Cooperation among these key participants is vital for achieving improvements in the energy efficiency of vessels and reducing the environmental impact of the maritime sector. Only then can the IMO strategy of carbon neutrality by 2050 be realised within the time frame.

10.5 SUMMARY

In this chapter, we have discussed the various projects spearheaded by the IMO to motivate and financially help the LDCs and SIDS to be part of the

IMO GHG strategy. Many shipping companies are already on the path to minimising the carbon footprint of their ships in an effort towards total decarbonisation. Their efforts have been discussed in some detail. Further, it is not only the ship-owners but also other stakeholders, such as flag administrations, PSC, classification societies, port authorities, and even ship designers, who have a role to play in the overall objective to reduce the emission of GHG. Their efforts have been discussed in order to get a holistic view of the efforts of the shipping industry in general for net neutrality. In the real-world context, it is important to know about these topics for several reasons, including environmental concerns, regulatory compliance, economic implications, and social responsibility.

In the next chapter, we shall discuss the efforts of the shipping industry beyond compliance requirements. Ship-owners have realised that they have to get ahead of the curve, and that is the reason they are embracing several innovative solutions towards achieving zero-carbon compliance well before the required timeline.

BIBLIOGRAPHY

54. IMO Report on the GreenVoyage2050 Project. https://greenvoyage2050.imo.org/about-the-project/.
    This report outlines the Project of IMO including its aims and the various components.
55. IMO Report on GHG SMART. (2019). https://www.imo.org/en/OurWork/PartnershipsProjects/Pages/GHG-Smart.aspx.
    This report gives details of the GHG SMART with details of the programme and the training package in order to train the stakeholders on the IMO GHG strategy and its implementation.
56. IMO Resolution MEPC 323(74). (Adopted on 17 May 2019). https://www cdn.imo.org/localresources/en/OurWork/Environment/Documents/Resolul tion323(74).pdf.
    This IMO resolution is an invitation to member states to encourage voluntary cooperation between port and the shipping sector with the aim of reducing GHG emissions.
57. Partnerships and Projects Between IMO and Stakeholders. (2019). https://www.imo.org/en/OurWork/PartnershipsProjects/Pages/default.aspx.
    IMO has initiated many partnership and projects, the details of which can be found here.
58. Green Awards Holders. https://www.greenaward.org/sea-shipping/incentive-providers/list-of-incentive-providers/.
    Details of ports and other organisations that provide incentives to holders of Green Award holders.
59. FAQs on Clean Shipping Index. https://www.dnv.com/maritime/advisory/csi-clean-shipping-index/faq/.
    These FAQs by the leading Classification Society DNV contain the answers to questions regarding CSI.

Chapter 11

Beyond Compliance
Surpassing Regulations, Carbon Credits, and Carbon Exchanges

We have seen the various methods and processes whereby a ship can remain in compliance with the different regulations and requirements related to energy efficiency. However, the shipping industry is now waking up to the fact that their ships must go beyond compliance if they are to meet the tough targets set by the International Maritime Organization (IMO) greenhouse gas (GHG) strategy.

Moving beyond mere compliance in the realm of energy-efficient shipping involves adopting proactive and innovative approaches that surpass the minimum regulatory requirements.

Incorporating cutting-edge technologies and implementing advanced operational practices, with a culture of continuous improvement, is the only way towards compliance and beyond.

In this chapter, we will discuss how the industry is moving beyond compliance in its endeavour to reduce GHG emissions and ultimately aim for zero-carbon shipping. The various technological developments and innovative ships being built are discussed. This will enable you to understand how the shipping industry is gearing up for the future.

First, let's look at why it's necessary to go beyond compliance.

11.1 WHY GO BEYOND COMPLIANCE?

The reasons why ship-owners would prefer to go beyond compliance in energy efficiency are many:

- Energy-efficient practices and operations would result in a reduction in fuel costs. Since fuel costs are a major portion of a ship's operational costs, any reduction in fuel consumption will directly lead to cost savings for the ship-owners and operators.
- Market positioning is another reason why the ship-owners may prefer to put procedures for energy efficiency over and above compliance. In today's competitive world, any advantage is seen favourably. The shipping industry has put the spotlight on sustainability and ships with

documented energy efficiency programmes will enjoy a competitive advantage over others.

- Many companies maintain a corporate social responsibility (CSR), now known as environmental, social, and governance (ESG), profile to attract potential charterers and other customers. To maintain their ESG activities, these shipping companies invest in advanced technologies to reduce their fuel consumption.
- Global warming and climate change are the headline stories these days. Hence, working in a company with a defined culture of sustainability will boost employee morale and will lead to the motivation of the employees towards a positive corporate culture. Further, this will also attract future talent and reduce attrition of employees.

Next, let us have a look at the ways and means that companies can utilise to remain ahead of the curve and move beyond compliance:

11.2 HOW TO GO BEYOND COMPLIANCE?

Going beyond compliance signifies a commitment to sustainability and a pursuit of excellence in increasing energy efficiency. In the following sections, we will look at some of the methods that are being adopted to achieve this.

11.2.1 Use of Technology

Energy efficiency technologies for ships are for reducing fuel consumption, emissions, and operational costs while enhancing sustainability in the maritime industry.

Shipyards and engine manufacturers are constantly exploring the use of innovative technology to increase energy efficiency on ships.

The European Commission conducted a study on energy efficiency technologies for ships.[60]

These technologies include hull design, propeller innovations, engine makeovers, and other innovative improvements to ship design and operations.

11.2.2 Ship Design

The design of a ship has a major role in reducing drag, frictional losses, and turbulence. All of these increase the energy efficiency of the ship. The bow of the ship plays an important part as it gives the wave-making resistance of the ship wherein a part of the energy generated by the propeller is utilised in creating the bow wave. Essentially this is a wastage of energy, and that is the reason the bulbous bow came into existence. Research is ongoing regarding the optimum size and shape of the bulbous bow. Companies like NYK and Maersk Line have modified the shape of their bulbous bow based on such research. It has been established that hull design is one of the major options to reduce fuel consumption. In fact, DNV-GL offers the innovative "Eco Lines", which streamlines the hull and increases fuel efficiency. Sophisticated software technologies such as computational fluid dynamics are often used to optimise the hull design for maximum streamlining and minimum turbulence.

11.2.3 Air Lubrication

This is another example of modern technology coming into play to increase the fuel efficiency of ships. The Mitsubishi Air Lubrication System (MALS) was one of the first to use this technology on the ship's hull. According to Wartsila, the use of air lubrication can reduce emissions by almost 10%. These systems are ideal for new buildings but can also be retrofitted on existing ships. The system works by releasing micro air bubbles from slots at the bottom of the ship. These air bubbles reduce the frictional resistance between the hull and the seawater resulting in lowering fuel consumption. Air lubrication systems have been around for over a decade, but it is only now after the GHG emission standards came into force that ship-owners are beginning to appreciate the advantage of such systems to reduce fuel consumption and comply with the EEXI and CII requirements. Air lubrication technology[61] is recognised by IMO as an "innovative energy efficiency technology".

11.2.4 Shaft Generator

The concept of shaft generators is not new. They take power from the main engine to produce electricity. Essentially shaft generators do not result in fuel efficiency as the energy required to produce electricity through these machines is the same as that through separate diesel generators. Many ship-owners opt for shaft generators as there is no need for diesel generators and thus there is a cost-benefit.

Engine makers have been working on the shaft generators to make them lighter and more fuel-efficient. With their improved functionality, these new design shaft generators result in substantial reductions in fuel consumption and, subsequently, increase the energy efficiency of the ships.

As stated by the engine maker ABB, development in power electronics technology, generator design, including permanent magnet machines, and higher performance control systems are being incorporated in new engines, resulting in increased fuel efficiency.

11.2.5 Propulsion Systems

Major engine makers are continuously working on propulsion systems to make them more efficient and reduce turbulence. There are many energy-saving devices (ESDs) currently available to streamline propeller systems, resulting in increased energy efficiency. These include the following:

- Pre-shrouded vanes (PSV) generate a pre-swirl flow in front of the propeller. This corrects the flow of water into the propeller and thus reduces the rotational energy loss of the propeller.
- Hub vortex absorbed fins (HVAF) or propeller boss cap fins (PBCF) are attachments to the hub of the propeller. This generates counter swirls to offset the propeller swirls. This breaks up the hub vortex and results in increased efficiency of the propeller.
- Wartsila, which is among the pioneers in improving the propeller design of existing ships, uses computational fluid design (CFD) to analyse the performance of the current propeller and the interaction between the propeller and the hull. This new propeller design results in improved propulsive efficiency.

There are other such design advances by the engine makers so that vessels are able to comply with the EEXI requirements and the CII ratings.

Many companies have taken the initiative to invest in technologically advanced ships which depend on alternate fuels and renewable energies. The catchword now is ESG, that is, environmental, social, and governance. This is an improvement on the earlier CSR (corporate social responsibility). By shifting to ESG, companies have shifted their focus to protecting the environment and sustainable development. Let us have a look at the developments of the future and the ships that are at the forefront of such development.

11.3 SHIPS OF THE FUTURE

Many companies are looking beyond compliance and designing innovative ships that will help usher in the zero-carbon era of shipping earlier than the IMO target dates. These vessels incorporate various technologies aimed at avoiding fossil fuel usage and minimising GHG emissions. The following are some of these companies and their innovative and futuristic ships.

11.3.1 NYK Line

NYK Line has been at the forefront of promoting environmental sustainability within their group. In March 2023 they established the NYK Group's environmental vision to support sustainable development by strengthening their environmental management activities. They have designed and are in the process of developing carbon-neutral ships. These are as follows:

- **Super Eco Ship 2030:** In 2009, the *NYK Super Eco Ship 2030* was designed with the aim of achieving an impressive 70% reduction in CO_2

emissions compared to current vessels due to its innovative design. This groundbreaking ship will utilise liquefied natural gas (LNG) fuel cells, emitting 30% less CO_2 than traditional marine diesel counterparts. Furthermore, the *NYK Super Eco Ship 2030* will feature an expansive 31,000 square metres of solar panels and retractable sails. The company has ambitious plans to have this environmentally friendly vessel operational by 2030. Noteworthy is the ship's inclusion of four pairs of sails to effectively harness wind energy. NYK has also outlined its long-term goal of achieving zero emissions across its fleet by 2050.

- **Super Eco Ship 2050:**[62] With the evolution of advanced technology and an increase in the IMO ambitions for reducing CO_2 emissions, building up to carbon-neutral shipping by 2050, NYK designed the *Super Eco Ship 2050* in 2018. Many innovations in energy utilisation, hull form propeller design, and machine technology have all resulted in the *Super Eco Ship 2050*, which boasts a 100% reduction in CO_2 emissions. Hydrogen fuel cells provide the energy for the propulsion of the ship while flapping foils, akin to the movement of dolphins, have replaced the traditional propellers. Renewable energy, such as solar power and wind energy, will be installed, further increasing the overall energy efficiency of the ship. Waste cold (the cold emanating from cold sources such as hydrogen, LNG, etc.) emanating from the liquid hydrogen bunker will be used for air conditioning and refrigerating. Conventional energy-saving practices, such as voyage optimisation, weather routing, and just-in-time arrival, will increase the energy efficiency of the ship.

11.3.2 AP Moller–Maersk (Maersk Line)

MAERSK Line has been striving to reduce GHG emissions from their ships by various means. Their aim is to implement a comprehensive ESG strategy with a commitment to reach net zero across our business by 2040.

Towards this end, they have ordered several dual-fuel ships capable of operating on green methanol in 2021. Recently, Maersk announced that they intend to retrofit an existing ship to a dual-fuel methanol-powered one in 2024. This project will also be replicated on sister vessels when they conduct their special surveys. Their aim is to retrofit the majority of their ships so that they can achieve net-zero emissions by 2040. This is a first in the industry and will pave the way for further such conversions from other ships. Retrofitting is a massive leap in reducing the carbon emissions of existing ships. If this trend catches on, the IMO strategy will be realised well ahead of their timetable.

They recently took delivery of the world's first large methanol-enabled container vessel in February 2024. It is the first of 18 such methanol-enabled vessels which will be delivered within a couple of years. These ships are built at HD Hyundai Heavy Industries (HD HHI) in Ulsan, South Korea.

These ships are 349 metres long with a capacity of 16,000 TEUs (20-foot equivalent).

According to Leonardo Sonzio, Head of Fleet Management and Technology at Maersk, "they wish to pave the way for future scalable retrofit programs in the industry and thereby accelerate the transition from fossil fuels to green fuels". Their overall goal is to demonstrate that methanol retrofits can be a viable alternative to new buildings.

11.3.3 Bernhard Schulte Shipmanagement

Bernhard Schulte Shipmanagement has been promoting environmental sustainability throughout their group for many years. They have established an initiative called BlueSeasMatter in order to encourage the shore and ship staff to come together for the protection of the environment and reduce the plastics in the oceans.

Carbon capture is one of the latest developments in the fight against CO_2 emissions. To promote this, BSM has entered the field of CO_2 transportation. In December 2023, they placed an order for their first CO_2 tanker in connection with the Northern Lights carbon capture and storage project. The order was placed with the Dalian Shipbuilding Offshore China (DSOC) and is scheduled for delivery in 2026. With the stringent regulations for GHG emissions coming in very shortly, carbon capture is seen as the way forward. Four such ships have been ordered, each capable of carrying 7,500 cubic metres of liquefied CO_2.

What the industry now needs are more ships to transport this captured carbon to the designated destination. Hence, there is a need for CO_2 tankers to transport the captured CO_2 from industrial sources. Some shipping companies are investing in these tankers in order to make it easier for CO_2 emitters to dispose of their CO_2. This will be a big boost to carbon capturing and storage (CCS). There is no doubt that CCS will become popular over the years. Thus, if ships are not available for transporting the captured CO_2, there will be a bottleneck. To avoid this and to prepare for the future, more such ships need to be ordered by existing shipping companies.

11.3.4 Wallenius Wilhelmsen

This is another company which takes environmental sustainability seriously. To facilitate this, they have adopted the "lean green sustainability strategy", which is in practice at all their ship and shore activities.

The Orcelle Wind, a wind-driven Pure Car Truck Carrier, is 220 metres long with a capacity exceeding 7,000 cars. Additionally, it will possess the capability to transport breakbulk and equipment or other cargo on wheels. This vessel holds a pivotal role in Wallenius Wilhelmsen's aim to be carbon neutral before IMO's timeline of 2050. It is scheduled to embark on its maiden voyage in late 2026 or early 2027.

Representing the inaugural ship from the Oceanbird concept,[63] which is primarily focused on wind-powered vessels (refer to the link given later), Orcelle Wind is at the forefront of showcasing the potential to slash emissions from vessels by up to 90 by various means of control of emissions.

When the Oceanbird concept is brought into the reality of commercial application, there will be challenges. This includes dependence on the wind force and direction for the speed of the ship, which may result in an inability to maintain ETA, dependence on traditional fuels for manoeuvring, maintenance issues, and so on. These challenges have to be worked out before the Oceanbird concept becomes acceptable to the commercial fleet.

In the global effort to tackle climate change, various solutions have emerged as a key tool for regulating GHG emissions across various industries, including the shipping industry. Shipping companies are key to this effort as they form an integral part of the global industry. Let us discuss some measures which can be considered as solutions to the problem of global warming.

11.4 ENERGY-EFFICIENT SOLUTIONS BEYOND COMPLIANCE

Global efforts to combat GHG emissions have a direct effect on shipping companies. For instance, the United Nations sustainable goals apply uniformly to all industries, and it is up to the industry regulators to adopt practices that foster sustainable development. Similarly, IMO has been consistent in its approach to reducing the carbon footprint of the shipping industry and has received full support from the stakeholders. Some schemes, such as emission trading, carbon capture, and so on, are not yet compulsory. These can be beneficial in many ways.

- Energy efficient solutions help to reduce GHG emissions. By going beyond compliance, businesses and individuals can contribute by lowering their carbon footprint.
- Cost savings are an important aspect of energy efficiency. Going beyond compliance can lead to even more cost savings and is beneficial in the long run.
- Adopting energy efficient solutions that go beyond compliance can position the business as a leader and thus gain competitive advantage in the market.
- Organisations are adopting the ESG principles. By going beyond compliance in energy efficiency contributes to environmental sustainability, have positive social impacts by lessening air pollution, and reflect strong governance practices.

Going beyond compliance is a voluntary effort on the part of businesses and is not only beneficial to their interests but vital for enhancing environmental sustainability.

11.4.1 Emission Trading Scheme

An emissions trading system (ETS)[64], also known as a cap-and-trade system, is a market-based approach to controlling and reducing GHG emissions. It is also known as the carbon trading scheme. It is designed to incentivise companies and industries to limit their carbon emissions by putting a cap, or limit, on the total amount of emissions that can be released into the atmosphere. The emission trading scheme will encourage shipping companies to invest in cleaner technologies rather than pay hefty sums in the carbon markets for purchasing carbon credits to offset their carbon emissions.

Established by the Kyoto Protocol, the scheme requires the government regulatory authorities to set a limit on CO_2 emissions, based on the industry. This limit differs for different industries and is based on transparent criteria. One carbon credit represents one metric tonne of carbon dioxide. Each entity is allowed to emit a certain amount of the pollutant known as allowance. The entity is expected to limit its emissions within the allowance allocated to them. An entity that emits fewer pollutants than its allocation can sell its excess allowance to those entities that are unable to limit their emissions within their allocation.

Emission trading is regulated by market dynamics. At the end of the compliance period, all the entities have to submit a record of their allowance versus their actual emissions, as well as any allowance sold or purchased by them from the market. If an entity has exceeded its allowance but could not buy from the market, it may be penalised by the authorities.

All countries have to set up their own trading scheme so that it acts as an incentive for entities to limit their emissions within their allowable allowance. These markets can be interlinked by bilateral agreements between countries or as part of international blocs, such as the European Union (EU) ETS. The trading scheme has been successful in that it acts as a deterrent for polluters, but its disadvantage is that it does not reduce the overall emissions. To counter this, some administrations have set up schemes where the administration themselves pick up excess carbon credits in order to lower these emissions.

From 1 January 2024, shipping has been included in the EU ETS. This is a major move as it addresses emissions from a major sector and provides an incentive for shipping companies to reduce emissions.

Another solution to curb GHG emissions across industries is carbon capture. Its use on ships can permit the ship-owner to use any type of fuel since the carbon dioxide will be captured and not released into the atmosphere. Let us delve into the system of carbon capture and storage and how it can be used on ships.

11.4.2 Carbon Capture and Storage on Ships

Carbon capture and storage (CCS)[65] refers to capturing CO_2 from the emissions from industries, such as power plants, as well as the steel and metal industries. CO_2 capture technology is not new to the industrial sector. As per the IEA, at the end of 2012, five large-scale CCS projects were operating around the world.

Carbon capture is not a requirement for ships as the logistics of this have to be worked out before it can be installed on board. Ship-owners continue to study the system to check whether it is feasible for their ships. Thus, we can say that the carbon capture and storage (CCS) is beyond compliance.

Carbon capture involves collecting the CO_2 emitted from various processes, such as thermal power plants and other industries. This CO_2 is then compressed and cooled to a liquid state and stored in pressurised vessels, ready for transportation. This liquefied CO_2 can be stored indefinitely in deep geological formations, discarded oil wells and lines, and even in unminable coal beds. This sort of carbon capture has been successfully conducted for years leading many companies to limit their CO_2 emissions.

Carbon capture on board is an exciting new method for the reduction of GHG from ships. Perhaps, in the long run, this may be a workable solution for reducing and eventually eliminating CO_2 emissions. However, there are many challenges for on-board carbon capture:

- The greatest challenge is the infrastructural requirements for capturing the CO_2 and then cooling it. This requires pre-treatment of the exhaust gases and is not an easy procedure because of the high temperature of exhaust gases. Ships have limited storage space, and the space required to store the captured carbon, even if compressed and liquefied, is mind-boggling.
- For capturing and storing CO_2, all non-CO_2 pollutants must be removed before compressing and liquefying the CO_2. This is another technological leap and requires additional infrastructural requirements.
- The pricing is another major challenge. The cost of carbon capture, storage, and transportation is prohibitive compared to the use of fossil fuels. This is one of the reasons why fossil fuels will continue to be used as long as the rules permit.

Unless CCS is made suitable for shipboard use, it will not emerge as a commonly used alternative for the reduction of GHG emissions from ships. Many companies are developing systems to make CCS affordable for shipboard use. Only then will this technology be a serious option and be acceptable for on-board use resulting in a substantial reduction of CO_2 emissions. Classification societies such as PNV and Bureau Veritas (BV) offer services to assist the ship operators in installing such systems on board.[66]

11.4.3 Carbon Sequestration

We have seen that CCS is being increasingly used in various industries to ensure that CO_2 is not released into the atmosphere. Carbon sequestration has emerged as another promising solution. It involves the entire cycle of capture of carbon both naturally and intentionally.

Carbon sequestration relates to the entire process of removing carbon dioxide from the atmosphere naturally (biological) or artificially as in CCS. It can be of various different types:

- **Biological Carbon Sequestration:** This is the natural storage of the carbon in carbon sinks. These may be vegetation such as forests, grasslands, oceans, or soil. It is a natural process, and this is one of the reasons we have to look after our forests and grasslands and take up afforestation wherever possible.
- **Geological Carbon Sequestration:** This is the term used when the captured carbon is stored in rock formations. Industries producing high levels of CO_2 such as power plants or cement factories resort to this method of injecting the captured CO_2 into porous rocks. This stores the CO_2 for a long term and effectively removes it from the atmosphere.
- **Industrial Carbon Sequestration:** This process involves the recycling of CO_2 and is the most economical and environment-friendly method of carbon sequestration. It has industrial uses, such as graphene production. Graphene is used for producing screens for technological devices, such as smartphones. The production of graphene requires a large amount of CO_2. It is also required in the chemical industry as well as in other industrial processes. The captured CO_2 is supplied to these industries, thereby preventing their escape into the atmosphere. However, the uptake of CO_2 by these industries is small compared to the overall emissions.

The Global Status of CCS Report 2023[67] by Global CCS contains the milestones for carbon capture and storage for the year.

So far, we have seen the various regulations governing GHG emissions from ships in order to curb global warming. However, there are initiatives over and above these regulations which can help in energy efficiency and reduce fuel consumption. Let's take a quick look at these.

11.5 LEGISLATION BEYOND COMPLIANCE

The International Organisation for Standardisation (ISO) has several standards that are specifically related to energy management and efficiency, the most notable of which are ISO 50001: Energy Management Systems (EnMS) and ISO 14001: Environmental Management Systems. These are environment standards that are not regulatory requirements. However, companies implement these standards in order to ensure that they are environment friendly and energy efficient. This helps them financially as fuel costs will reduce, and commercially as they can then build an image for themselves in the market. The difference between these two standards is that while ISO 14001 identifies and manages environmental aspects and impacts, ISO 50001 focuses on improving energy performance, reducing energy consumption, and enhancing energy efficiency. The Poseidon Principles, which are increasingly being adopted by banks and marine insurers, are another example of legislation beyond compliance. Let's look at them in greater detail.

11.5.1 ISO 14001

ISO 14001 is a standard for environment management systems, the current version of which is ISO 14001–2015. This standard is often used by ship-owners and operators and is adapted for the shipping industry by classification societies who have developed and maintained their quality management system ISO 9001–2015.

ISO 14001 for ships is tailored for the maritime industry and ensures that the ship as well as the company meets all the required criteria for environment management. Since this is not a mandatory requirement, we can say that it is beyond compliance and will assist the ship to further reduce GHG emissions. Regular ISO 14001–2015 audits will help the ship-owner or operator to identify shortfalls in the environment management system and take corrective actions to address the non-conformities, if any identified during the audits.

Companies must have an environmental policy reflecting the commitment of the leadership to prevent pollution, continuously improve environmental pollution, and comply with legal requirements. Classification societies accredited by the administration can issue these certificates to ships after ensuring that the ships comply with the requirements and have successfully cleared the audits without any outstanding non-conformities.

Implementing ISO 14001 on boards ships and in the shore office can substantially enhance energy efficiency. Like all other ISO standards, continuous improvement is the key word, wherein the ship operators and the ships will be encouraged to utilise their resources effectively in order to reduce fuel consumption. This will result in regulatory compliance and cost benefits.

11.5.2 ISO 50001

This is a standard related to energy management systems applicable to various industries including the shipping sector. It is aimed at organisations to help them manage their energy use and improve their energy performance.

ISO 50001 is a general standard and not specifically tailored for ships, but the principles of the standards can be applied to address energy-related challenges in the future.

Classification societies accredited by the administration usually sets up the energy management system as per the requirements. They can then issue ISO 50001 certification to those companies who are able to provide evidence of compliance with the energy management system and successfully clear the required audits.

This standard can be voluntarily implemented both on board and in the shore offices. The basic purpose of ISO 50001 on board ships is to streamline energy management and improve energy efficiency while ISO 14001 is targeted towards environmental management systems, including waste management. While both the standards can be used for the ultimate aim of improving energy efficiency, it is left to the ship operators to implement either one or even both standards on board the ship in order to encourage sustainable operations.

11.5.3 ISO 14064

This is a standard for measuring and reporting GHG emissions accurately and transparently. The integration of ISO 14064 within a shipping company and across its fleet can play a pivotal role in ensuring adherence to IMO regulations concerning the DCS and the DCP. This is achieved through the establishment of a standardised system for GHG accounting, data handling, verification, and ongoing enhancement. This will ensure transparency, accountability, and trustworthiness in the reporting of emissions, thereby bolstering the sustainability of the maritime sector.

This is also a voluntary standard that can be implemented by ship operators on board which will help in the accurate measurement and reporting of GHG emission data.

11.5.4 The Poseidon Principles

The Poseidon Principles[68] represent a groundbreaking initiative within the maritime industry, aimed at aligning financing practices with global climate goals. Introduced in June 2019, this is a global framework endorsed by prominent banks, shipping entities, and other stakeholders in the maritime domain. They integrate climate considerations into the decision-making processes of financial institutions, insurers, and other stakeholders regarding lending within the shipping sector. Named after Poseidon, the Greek deity symbolising the seas, these principles underscore a commitment to fostering oceanic sustainability.

The Poseidon Principles are an important step towards sustainable development. For example, financial institutions may evaluate the impact of carbon pricing and emission regulations before investing in shipping companies or extending loans to them. Thus, this framework will encourage environmentally sustainable practices and investment in clean technologies. Similarly, major insurers may look at the GHG performance of ships before deciding on the premium.

Adherence to the Poseidon Principles will ensure that ship-owners migrate to zero-carbon options and seriously think about the use of alternate fuels and renewable energy for powering their ships.

11.6 SUMMARY

In this chapter, we have seen how shipping companies and stakeholders can go beyond compliance in order to reduce GHG emissions. Many ship-owners are in the process of designing carbon neutral ships, relying solely on alternative fuels or renewable sources of energy. We have also discussed the importance of emission trading schemes as an incentive to adopt carbon neutral technologies. We then looked at carbon capture, which is the technology of the future and will soon be adopted by major shipping companies as an alternative to CO_2 emissions once it becomes cost-effective. Finally, we discussed legislative actions that go beyond compliance, such as ISO 14001, 14064 and 50001, which are standards for environment management. The shipping industry has further responded by voluntarily including the Poseidon Principles in their financial transactions and insurance contracts. These topics are important in the real world because of the ultimate aim to reduce GHG emissions and consequently, global warming. In addition, the cost savings that accrue with energy efficiency, the need for regulatory compliance, and so on are important benefits of the efforts towards energy efficiency which will ultimately lead to a sustainable future.

In the next chapter, we shall discuss the best practices for fuel efficient operations of ships. We shall discuss how ships can become more fuel efficient resulting in the reduction of GHG emissions.

BIBLIOGRAPHY

60. Study on Energy Efficiency Technologies for Ships. http://publications.europa. eu/resource/cellar/302ae48e-f984-45c3-a1c0-7c82efb92661.0001.01/DOC_1.

    This study conducted by the European Commission highlights the various uses of technology to improve the energy efficiency on ships.
61. Air Lubrication Technology. (2019). https://ww2.eagle.org/content/dam/eagle/ advisories-and-debriefs/Air%20Lubrication%20Technology.pdf.

    The concept of this technology as well as its advantages have been explained. The impact on EEDI is also shown.
62. NYK *Super Eco Ship 2050*. https://www.nyk.com/english/esg/pdf/ecoship2050_ en.pdf.

    The *NYK Super Echo Ship* is innovatively designed with modern technology such as hydrogen fuel battery, WHR, solar power, air lubrication system and so on.

63. The Oceanbird Concept. https://www.theoceanbird.com/the-oceanbird-concept/.
    The Oceanbird is a concept to achieve zero-carbon ships by 2027. The
    Orcelle would be the first vessel from this concept, being the world's first wind-
    powered RoRo vessel.

64. IEA. (2020). *Implementing Effective Emission Trading Systems.* Paris: IEA.
    https://www.iea.org/reports/implementing-effective-emissions-trading-sys
    tems. Licence: CC BY 4.0.
    Emission trading systems is the method to incentivise energy efficiency and
    IEA has issued this report on implementation of the same.

65. IEA Report on Carbon Capture System. (2023). https://www.iea.org/
    energy-system/carbon-capture-utilisation-and-storage#how-does-ccus-work.
    This report by the IEA comprehensively explains carbon capture and stor-
    age, their role in energy transition as well as the technological innovations in
    this field.

66. CCUs: The Overlooked Technology Behind the Marine Energy Transition. (2024).
    https://marine-offshore.bureauveritas.com/ccus-overlooked-technology-
    behind-marine-energy-transition.
    There is no doubt that carbon capture would be one of the best solutions in
    eliminating GHG emissions from ships. This report by BV explores the benefits
    and the process of carbon capture from ships and their utilisation and storage.

67. Milestones for Carbon Capture and Storage. (2024). https://status23.globalc
    csinstitute.com/.
    This report by the Global CCS Institute progress and development of car-
    bon capture across the world. As per the report, there is good progress in CO_2
    capture as well as the new number of facilities being added.

68. The Poseidon Principles. (2023). https://www.poseidonprinciples.org/insurance/
    #home.
    This homepage of Poseidon Principles contains the details of the principles.
    The Poseidon Principles themselves can be downloaded if required.

Chapter 12

Guidance on Best Practices for Fuel-Efficient Operations of Ships

The shipping industry has devised the best practices for increasing the fuel efficiency of ships. Many associations, such as the International Association of Independent Tanker Owners INTERTANKO), the Baltic and International Maritime Council (BIMCO), the International Association of Classification Societies (IACS), the United Nations Framework Convention on Climate Change (UNFCC), and even the European Union, have released some best practices for ensuring ships' fuel efficiency. Although they are all broadly aligned to the common goal of International Maritime Organization (IMO) towards reducing greenhouse gas (GHG) emissions, there are some differences between them. In this chapter, we shall have a look at these best practices, as recommended by the various stakeholders, to understand their intricacies.

It is abundantly clear that one of the prime measures to reduce GHG emissions is to reduce the consumption of fuel. A case in point is that of a multinational company that was blacklisted by the flag administration in early 2023 for its failure to implement energy-saving measures on board its ships during a routine flag state inspection. The flag administration gave them six months to show tangible results in increasing energy efficiency. The company took up the matter seriously, trained their senior officers in energy efficient techniques, and implemented various corrective actions based on energy audits carried out on board. This resulted in a marked increase in the fuel efficiency of their ships and of the fleet as a whole. The commitment of top management is no doubt one of the primary requirements in implementing the best practices for reducing the emission of GHG from ships.

There is no doubt that the best method to reduce GHG emissions is using alternate fuels and investing in technology for newer equipment. On board, there are many operational practices that can be followed to achieve energy efficiency. Let us first discuss the IMO guidance for fuel efficiency on ships.

Implementing best practices and guidelines is the keystone for improving energy efficiency on ships. This the reason that many international organisations including IMO issues guidelines and best practices to be followed on board ships. Ship operators make a careful note of these guidelines and implement them on board to improve energy efficiency and sustainability.

DOI: 10.1201/9781032702568-12

12.1 IMO GUIDANCE

IMO has been constantly working to find out solutions to the problem of reduction of GHG emissions.[69]

The IMO Resolution 346(78)[70] adopted on 10 June 2022 deals with the guidelines for the development of a Ship Energy Efficiency Management Plan (SEEMP). This resolution also provides guidelines to help implement best practices for fuel-efficient operations. Let's look at some of these in detail.

12.1.1 Improved Voyage Planning

According to this guideline, "The optimum route and improved efficiency can be achieved through the careful planning and execution of voyages." Voyage planning should be the safest and most economical route from one port to the other. The effort should be to take the shortest route since this will have a direct correlation to the distance covered and the fuel consumption. At the same time, safety should never be compromised. The voyage plan is prepared by the navigating officers and approved by the master of the ship. Many accidents have taken place where the voyage plan has been prepared without due care. IMO resolution A.893(21) provides the guidelines for voyage planning and is to be referred to.

On 5 October 2011, the container vessel *Rena* ran aground on the Astrolabe reef near Tauranga, New Zealand. The grounding and capsized resulted in a massive oil spill of more than 300 tonnes of heavy fuel oil and caused an environmental disaster by damaging the coral reef the flora and fauna, as well as the pristine beached of Tauranga. It occurred due to the master taking a shortcut in order to maintain the time of arrival at the next port. This accident underlines the importance of safety over commercial aspects during the voyage.

12.1.2 Weather Routing

This guideline states, "Weather routeing has a high potential for efficiency savings on specific routes. It is commercially available for all types of ship and for many trade areas." Weather routing is used to optimise the route of a vessel based on weather conditions. In earlier days, the master of the ship was responsible for navigating the ship in the safest way, keeping in mind the weather that may be encountered during the voyage. With the advent of satellite communication and advanced weather monitoring capabilities, most ship-owners outsource weather routing to professionals, who utilise sophisticated technology to advise the master on the best route. Taking advantage of favourable wind and currents and avoiding storms and areas of bad weather are some of the advantages of weather routing for the ship. In addition, there is software available with advanced algorithms and satellite connectivity that can provide the ship with real-time weather data, enabling them to navigate safely and in the most fuel-efficient manner.

12.1.3 Just in Time

This guideline recommends, "Good early communication with the next port should be an aim in order to give maximum notice of berth availability and facilitate the use of optimum speed where port operational procedures support this approach." This is known as just-in-time arrival. Ships proceed at full speed as a normal procedure to reach the next port at the earliest. They then anchor and report to the port for berthing instructions. This may be the same day, the next day, or even after a few days. In order to reduce fuel consumption, there should be proper communication with the port so as to get a correct indication of berth availability. The ship can then proceed at optimum speed and reach well before the required time, reducing fuel consumption without endangering their port schedules.

12.1.4 Speed Optimisation

This guideline states the following:

> Speed optimization can produce significant savings. However, optimum speed means the speed at which the fuel used per tonne mile is at a minimum level for that voyage. It does not mean minimum speed; in fact, sailing at less than optimum speed will consume more fuel rather than less. Reference should be made to the engine manufacturer's power/consumption curve and the ship's propeller curve. Possible adverse consequences of slow-speed operation may include increased vibration and problems with soot deposits in combustion chambers and exhaust systems. These possible consequences should be taken into account.

Usually, we use the phrase "slow steaming" in order to reduce fuel consumption. Slow steaming does not mean minimum speed. There are two speeds we consider, full speed and reduced speed, also known as slow speed or optimum speed. For example, a ship that has a cruising speed of 18 knots can achieve a speed of 6 knots when going dead slow ahead, also known as minimum speed. But to obtain the best fuel efficiency as compared to the distance travelled (in tonne-mile), a speed of 12 knots or at half speed may be advisable. Thus, engine manufacturers usually provide reference power/consumption curves which should be taken into account when deciding to proceed at a slower speed to improve energy efficiency.

However, proceeding at a slow speed has certain disadvantages, which the chief engineer has to keep in mind:

- The first is keeping track of the revolutions per minute (RPM). The critical RPM of the engines refers to the RPM at which the rotational speed of the propeller shaft coincides with its natural frequency. When these two frequencies coincide, resonance occurs, leading to severe vibrations. Thus, when the ship proceeds at a slower speed, the RPM should be well above or below the critical RPM.

- Another disadvantage of proceeding at a slow speed is that soot particles which at normal RPM would have been carried out through the exhaust are deposited in the combustion chambers when the vessel proceeds at a slower speed. Soot deposits can be quite dangerous as they can result in fires in the scavenge spaces. Chief engineers of such ships overcome this by proceeding at full speed every few hours for a short while so that the deposited soot particles are blown away through the exhaust.

12.1.5 Optimised Shaft Power

According to this guideline, "Operation at constant shaft RPM can be more efficient than continuously adjusting speed through engine power. The use of automated engine management systems to control speed rather than relying on human intervention may be beneficial." Operating a ship at constant shaft RPM can be more efficient than continuously adjusting speed through engine power in certain situations. This concept is often associated with the idea of operating a ship within its "Economic Speed Range." It results in fuel consumption stability, optimal propeller and engine efficiency, and improved performance of the ship. Thus, adjusting speed through automated systems will increase fuel efficiency. However, in cases of adverse weather, specific operational requirements, or varying sea states it may be necessary to adjust speed through engine power.

12.1.6 Optimised Ship Handling

Optimised ship handling is one of the important on-board actions in order to reduce fuel consumption. The ship's officers need to understand these measures and put them into practice. These measures include the following:

- Optimum Trim:

  Loaded or unloaded, trim has a significant influence on the resistance of the ship through the water and optimising trim can deliver significant fuel savings. For any given draft there is a trim condition that gives minimum resistance. In some ships, it is possible to assess optimum trim conditions for fuel efficiency continuously throughout the voyage. Design or safety factors may preclude full use of trim optimisation.

  The hull resistance of the ship is largely dependent on its draft and trim. The draft of the vessel depends on the cargo loaded and cannot be controlled except for commercial reasons. However, it is possible to adjust the trim to the optimum required for better fuel efficiency.

The optimum trim varies from ship to ship and has to be determined based on the vessel's speed and draft. Consequently, by proper cargo planning, the trim can be optimised for fuel efficiency. There is equipment available for trim optimisation, such as a trim optimiser or a better loading computer, fitted with software that can calculate the required trim for fuel efficiency.

- **Optimum Ballast:**

Ballast should be adjusted taking into consideration the requirements to meet optimum trim and steering conditions and optimum ballast conditions achieved through good cargo planning. . . . Ballast conditions have a significant impact on steering conditions and autopilot settings, and it needs to be noted that less ballast water does not necessarily mean improved energy efficiency.

As a rule, the vessel should take the minimum ballast, keeping in mind the safety of the vessel and the weather conditions. It is common knowledge that the energy required to move a ship forward increases as the displacement of the ship increases. Thus, carrying more than the optimum ballast will lead to an increase in fuel consumption. At the same time, carrying less than the optimum ballast will affect the safety of the ship, especially in adverse conditions. The ship's Ballast Water Management Plan should always be referred to when planning the ballast.

- **Optimum Use of Rudder and Heading Control Systems (Autopilots):**

There have been large improvements in automated heading and steering control systems technology. Whilst originally developed to make the bridge team more effective, modern autopilots can achieve much more. An integrated Navigation and Command System can achieve significant fuel savings by simply reducing the distance sailed "off track". The principle is simple: better course control through less frequent and smaller corrections will minimize losses due to rudder resistance. Retrofitting of a more efficient autopilot to existing ships could be considered.

The setting of the rudder on autopilot should allow for the vessel to go off track to a certain limit. If not, there will be frequent course corrections applied by the autopilot which will reduce the speed of the ship and thereby the fuel efficiency. Many autopilot systems have a rough weather setting, which will reduce the rudder movements. However, the watchkeeping officer should realise the limitations of the autopilot in congested waterways and keep the vessel on hand steering in order to facilitate quick response.

12.1.7 Hull Maintenance

This guideline recommends the following:

> Docking intervals should be integrated with the company's ongoing assessment of ship performance. Hull resistance can be optimized by new technology-coating systems, possibly in combination with cleaning intervals. Regular in-water inspection of the condition of the hull is recommended. Propeller cleaning and polishing or even appropriate coating may significantly increase fuel efficiency. The need for ships to maintain efficiency through in-water hull cleaning should be recognized and facilitated by port States.

The most important point here is that regular underwater hull inspection should be carried out to determine the condition of the hull. Excessive growth on the ship's hull is known as biofouling and consists of algae, barnacles, mussels, and so on. To prevent this, anti-fouling paints are applied during the dry-docking of the ship, and regular maintenance of the hull is carried out.

In 2008, I joined a container vessel as master. On taking over the ship, I found that the speed of the vessel at full ahead was only 20 knots instead of the rated 24 knots. There was a lot of communication between the shipowners, ship managers, charterers, and so on regarding this issue. The next port was Singapore, and I requested an underwater inspection. This was carried out, and it came to light that the bottom of the ship was heavily fouled. Hull cleaning and propeller polishing were carried out immediately. During the next voyage, the speed picked up to 23 knots. This clearly shows the relation between hull maintenance and fuel efficiency. Hull fouling increases frictional losses and thereby fuel consumption.

12.1.8 Propulsion System Maintenance

This states:

> Maintenance in accordance with manufacturers' instructions in the company's planned maintenance schedule will also maintain efficiency. The use of engine condition monitoring can be a useful tool to maintain high efficiency. Additional means to improve engine efficiency might include use of fuel additives, adjustment of cylinder lubrication oil consumption, valve improvements, torque analysis, and automated engine monitoring systems.

Maintenance of the equipment and machinery is an important component of fuel efficiency. Some methods for ensuring adequate maintenance include the following:

- As per the requirements of the International Safety Management (ISM) Code, vessels must implement an approved PMS to ensure that all the machinery is maintained in a proper condition. Properly maintained machinery will improve efficiency and reduce fuel consumption.

- Engine condition monitoring is the process of monitoring the health of the engines by studying the engine parameters. An analysis of these parameters would give an idea of the health of the equipment and a warning that a potential malfunction may occur.

We have seen the best practices on board as per IMO Resolution MEPC 346(78). It also underlines the best practices recommended for ship-owners in order to increase fuel efficiency, such as enhanced propeller design, WHR, improved fleet management, improved cargo handling, the establishment of an energy management system, and so on. These topics have been discussed in the earlier chapters. The IMO guidelines cover the most important measures that can result in reducing GHG emissions.

Now let us take a quick look at some of the best practices over and above the IMO guidelines.

12.2 OTHER BEST PRACTICES

Further to the IMO guidelines and aligned with them, classification societies, shipping companies, flag administration, and so on have brought their own guidelines in the form of circulars, technical bulletins, and so on. Some of the more relevant such practices are as follows:

- Shipping companies are responsible for their ships and must have commitment, leadership, and vision if they are to ensure that their ships fully comply with the relevant regulations and reduce the emission of GHGs. They should formulate a policy to ensure fuel efficiency on their ships and monitor the performance to ensure that the policy is implemented and complied with. Management commitment is very important if the policy is to be effectively implemented and efforts are to be invested in increasing fuel efficiency and reducing fuel consumption.
- As the hull resistance increases, the fuel efficiency goes down. Thus, the company should implement a resistance management programme, including propeller resistance, hull resistance, and so on. Hull and propeller cleaning should be done regularly. A resistance management programme helps the company to monitor and analyse the various resistance factors affecting the ship. Various factors, such as biofouling, trim, and draft, affect the resistance of the vessel and increase frictional losses, leading to increased fuel consumption. The optimum trim, minimum ballast, regular hull cleaning, propeller polishing, and speed optimisation are some of the factors that are part of the resistance management programme. Monitoring these factors and taking corrective action as required will lead to reducing the resistance and thereby increasing fuel efficiency.

- Propulsion machinery is one of the pillars of fuel efficiency. There are many ways to increase the efficiency of the propulsion machinery, as discussed earlier. Machinery optimisation programme, main engine monitoring, and the optimisation of lubrication and other machinery and equipment can be implemented and followed on board in order to increase fuel efficiency.

  Machinery optimisation programme is the management of the ship's machinery systems in order to increase their efficiency and reduce fuel consumption. In order to increase the efficiency of the main engines and the auxiliary equipment, performance monitoring is important. Performance monitoring includes monitoring parameters such as exhaust temperatures, the temperatures and pressure levels, fuel consumption, and so on. There are automated systems for analysing these data and identifying deviations.

  Most companies install a PMS whereby the inspection and maintenance of the machinery at periodic intervals. If the PMS is properly carried out, the maintenance of the engines and associated equipment will improve, thereby enhancing their reliability and efficiency.

- Cargo handling is also a means of fuel inefficiency. This aspect of saving energy has not received the attention it deserves. In many cases, the cargo handling equipment is kept in running condition well before the start of cargo operations and even during breaks. Generally, these pumps and motors are heavy-duty equipment, using a lot of energy. Ensuring care in the operation of these equipment can help to reduce energy utilisation and thereby reduce fuel consumption. Proper maintenance of cargo handling equipment will increase fuel efficiency.

- Tankers have large cargo oil pumps (COP) for pumping the cargo ashore while discharging. If these pumps are not maintained satisfactorily or their operation is not controlled, the energy efficiency is reduced as these pumps utilise a large amount of energy. Many ship officers have a practice of recirculating the oil by means of these pumps due to various operational reasons. This practice should be discouraged on board in order to increase energy efficiency as these pumps consume a lot of energy.

- Mooring equipment on board is driven by powerful motors and pumps. In order to reduce energy loss, this equipment should be judiciously operated. For example, it is a common practice on ships to operate this equipment well before it is required, leading to a loss of energy. It should be made clear that such equipment should not be idling when not required.

- Air conditioning is another energy guzzler on the ship. The chief engineer of the ship is in charge and should ensure that the temperature in the cabins and public places should be comfortable and not cold. This is another energy-saving policy that is to be adopted by ships in their efforts at fuel efficiency.

- Crew awareness is an important aspect of increasing shipboard fuel efficiency. The crew either are not aware of the importance of energy efficiency or do not know the procedures for reducing energy loss on board. In order to overcome this, training in energy efficiency is required so that there is a holistic effort at reducing energy losses. The training should also include the various equipment considered critical to energy efficiency and detail their use in a fuel-efficient manner.
- The SEEMP generally contains all the methods and practices to increase fuel efficiency. The plan must be discussed on board with the crew and all relevant measures should be implemented. The monthly safety committee meeting is a good time to discuss the best practices for energy efficiency. The master must take a pro-active interest in this matter and explain to the officers and crew the need for fuel efficiency and reduction of fuel consumption.
- Keeping the above factors in mind companies must formulate a SOP detailing the controlled operation of these equipment. These SOPs should be clearly explained to the crew and posted in places adjacent to the equipment.
- An energy conservation awareness plan is to be implemented on board and onshore. This is a structured approach to implementing energy-saving practices within the company's operations, both at sea (on board) and on land (onshore). Companies develop this plan to ensure energy efficiency while also meeting regulatory requirements both on board and onshore. Personnel should be trained and familiarised with the company's energy efficiency programme, including an accommodation-specific energy conservation programme. It is only with the active involvement of all ship staff that the fuel efficiency of the vessel increases.

In general, these best practices are a guide to the ship-owners, operators, and ship staff on the way forward to achieve energy sustainability and compliance with regulations. Now let us have a look at the positions adopted by major industrial organisations related to energy efficiency. They are influential organisations that work on various aspects of energy efficiency, sustainability, and regulatory compliance.

12.3 INTERNATIONAL ASSOCIATION OF CLASSIFICATION SOCIETIES

This is an organisation consisting of 12 major classification societies. They are the nodal agency for classification societies who are loosely bound by their guiding principles. Their common structural rules (CSR) ensure uniformity in the application of the structural design of ships at the time of shipbuilding. Classification societies follow these rules so that their ships meet the structural integrity and strength as per the standards of shipbuilding.

IACS promotes energy efficiency and sustainability in the maritime industry by ensuring regulatory compliance, setting standards for compliance, research, training and education.

Classification societies are bound to follow these standards to ensure that the ships are energy efficient. As an example, the calculation of EEDI, EEXI and CII as mandated by IMO is a very complicated affair. Ship-owners and operators often face difficulty in understanding them. Classification societies have expertise in this field and offer technical support for the implementation of these standards. Right from the ship-building stage, they work closely with the shipyards to ensure that the ship is in compliance with these requirements.

For existing ships, they offer guidance and knowledge sharing to help the ship to comply with the EEXI and CII requirements. They conduct surveys on board ships and based on this suggest corrective actions such as energy efficient operations, adopting technological installations, and so on.

Both IACS and the individual classification societies offer the best practices for energy efficiency on ships. These best practices broadly follow the IMO guidelines. In addition, they bring out circulars and detailed guidelines on the methodology to implement these best practices. In short, they seek to ensure that the ships are aware of the requirements and thereby contribute to sustainable operations. This will lead to the compliance of the GHG strategy of IMO leading to reduction in GHG emissions and mitigate the global warming.

12.4 INTERTANKO'S GUIDELINES

The International Association of Independent Tanker Owners (INTERTANKO) is a prominent international industry association that represents the interests of independent tanker ship-owners and operators. It was established in 1970 and is based in Oslo, Norway. INTERTANKO's primary focus is on the tanker sector of the shipping industry. It takes an active role in the current scenario of climate change and reduction of GHG. It advises its members on the various aspects of energy efficiency and how to achieve the IMO targets of GHG emissions and fuel efficiency. INTERTANKO supports "ship-neutral" regulations. This means that the regulations should apply equally to all ships regardless of their flag state. This aligns with the IMO conventions, like MARPOL and SOLAS, which apply to all applicable ships.

INTERTANKO's position reflects a commitment to reducing GHG emissions from tankers and the shipping industry as a whole, while also emphasising the need for balanced regulations that consider both environmental goals and the practicalities of global commerce. It supports energy efficiency measures, international coordination, and ambitious yet achievable targets for emission reduction. To this end, it sends circulars, advisories, and technical bulletins to its members as a means of educating them and breaking down the complex regulations into simple words for ease of understanding. INTERTANKO offers practical guidelines to its members on enhancing energy efficiency on board their ships and thereby reducing GHG emissions. The gist of these guidelines is as follows:

- **Technical and Operational Guidance:** INTERTANKO provides technical circulars, manuals, guides, and other best practice documents on topics such as fuel management, engine performance optimisation, hull maintenance, and so on.
- **Knowledge Exchange:** It conducts seminars and encourages knowledge exchange through forums and working groups in order to discuss challenges and find solutions to the problem of global warming.
- **Research and Development:** INTERTANKO supports research and development in energy efficiency by funding research projects, forming technical working groups to identify such projects, and sharing research findings and best practices with its members.

INTERTANKO contributes to developing technology and best practices to enhance the fuel efficiency of ships and therefore reduce GHG emissions.

12.5 BIMCO GUIDELINES

The Baltic and International Maritime Council (BIMCO) is a global shipping association that plays a significant role in shaping the policies and standards within the maritime industry. It does not have a direct regulatory authority but provides guidelines, contracts, and best practices to its members, contributing to the efficient and sustainable operation of ships. BIMCO supports the efforts of the shipping industry in the reduction of GHG emissions. To this end, it advises its members on the best practices to increase fuel efficiency.

BIMCO's efforts at increasing energy efficiency are similar to the efforts of INTERTANKO. Its efforts to address the challenges of climate change by supporting research, providing education and guidance as well as through policy advocacy have demonstrated the commitment to such international bodies to promote environmental sustainability. They work tirelessly for the improvement of their members and global shipping in general. They fully support the IMO initiatives to reduce GHG emissions and continuously guide and advise their members on the latest requirements and the ways and means to implement them. They also issue best practices for the same which are broadly aligned with the IMO's recommendations on best practices to reduce GHG emissions.

BIMCO has developed several charter parties for various requirements, such as time charters, voyage charters, and so on. Keeping in mind the urgent need for energy efficiency and reduction of GHG emissions, BIMCO has come up with several clauses to ensure that vessels and ship operators comply with the relevant IMO regulations:

- The carbon intensity clause authorises ship operators and the master to alter course or reduce the speed of the vessel if this is required to reduce the carbon intensity of the vessel.
- The slow steaming clause empowers the owners to instruct the master to proceed at slow speed if required, provided the commercial interests are not compromised.
- The EEXI transition clause requires the vessel to comply with the EEXI requirements of IMO including the implementation of the engine power limitation, the shaft power limitation, the carbon intensity index, and so on.

The aim of these clauses is to ensure that the ship or her owners comply with the relevant regulations and contribute to sustainable operations.

12.6 SUMMARY

In this chapter, we discussed the importance of best practices. The IMO outlines the recommended practices regarding energy efficiency in MEPC Resolution 346(78). Beyond the IMO guidelines, some practical measures to reduce fuel consumption have been discussed, such as the operation of mooring equipment, cargo handling gear, and so on. Resistance management and machinery management are also tools that the ship-owner or operator can apply on board in order to reduce fuel consumption on board. Gaining an understanding of the best practices to be adopted in order to increase the fuel efficiency of ships is crucial for both shipping companies and the shipboard staff to follow the recommendations in both letter and spirit. We have also discussed the efforts of international organisations like INTERTANKO and BIMCO to reduce GHG emissions.

In the next chapter, we will discuss some important case studies, focusing on various incidents that resulted in legislative changes to curb pollution of the seas and the marine environment. Case studies are important because their investigation will lead to the root cause and the preventive action to avoid the repetition of such incidents.

BIBLIOGRAPHY

69. Improving the Energy Efficiency of Ships. https://www.imo.org/en/OurWork/Environment/Pages/Improving%20the%20energy%20efficiency%20of%20ships.aspx.

This report sheds light on the IMO's efforts to reduce the GHG emission from ships right from 2013 till 2023.

70. IMO Resolution 346(78). (Adopted on 10 June 2022). *Guidelines for the Development of a Ship Energy Efficient Management Plan (SEEMP)*. https://wwwcdn.imo.org/localresources/en/KnowledgeCentre/IndexofIMOResolutions/MEPCDocuments/MEPC.346(78).pdf.

Chapter 13

Case Studies on Environmental Pollution

Over the years, many oil spills have occurred across the world from both tankers and non-tankers. Spills from tankers are more damaging because of the quantity of oil spilled. A non-tanker has oil in the form of fuel, which is usually in a limited amount. But tankers carry large amounts of oil as cargo, and consequently, a spill is generally catastrophic.

In general, case studies on environmental pollution are important to understand the root cause of the pollution. Knowing and studying about oil spills from ships is essential for protecting the environment, safeguarding human health, sustaining economic activities, and promoting international cooperation in addressing marine pollution. In this chapter, we will use some popular case studies focused on major oil spills to help highlight the corrective and preventive actions taken and identify the patterns and trends of pollution. By studying such real-world examples, policymakers can develop and implement regulations and guidelines as preventive measures.

Through these case studies, we aim to understand the development of the International Convention for the Prevention of Marine Pollution (MARPOL). We will also discuss a fictional case study of a company that managed to go beyond compliance and achieved the International Maritime Organization's (IMO's) targets well ahead of the required dates. Let's start with a brief understanding of what the impacts of oil spills are.

13.1 IMPACTS OF OIL SPILLAGE

It is widely known that oil spills can have a serious impact on the environment, including the ecosystem and the economy, as well as on public health. Mangroves, coral reefs, coastal areas, and marine life in general can be seriously affected by oil pollution. Fish and birds cannot survive in oily waters and will perish.

However, air pollution from evaporating oil during an oil spill is often ignored or glossed over. The amount of evaporation depends on various factors. In their technical information paper, "Fate of Marine Oil Spills",[71] the International Tanker Owners Pollution Federation Limited (ITOPF) has

DOI: 10.1201/9781032702568-13

stated that the percentage of evaporation from an oil spill depends on many factors, such as the properties of the oil, the ambient temperature, the state of the sea, and the wind speed. Warm weather with a strong breeze will result in greater evaporation than in colder climes.

Evaporation also depends on the type of oil. Lighter crude oils, such as Cossack crude, evaporate at a higher rate than heavier crudes. Each variety of crude oil consists of oils of different boiling points. For instance, 55% of Cossack crude is formed of components that have boiling points below 200°C, and thus almost 55% of the oil spilled will evaporate within 24 hours. This is a huge amount and is the reason that response to oil spills must be immediate in order to capture the spilled oil before it evaporates.

Let us now have a look at some of the major oil spills across the globe in the last few decades.

13.2 MAJOR OIL SPILL INCIDENTS

As discussed in the previous section, oil spills result in various forms of marine pollution, including air pollution. Consequently, IMO brought in new regulations as a response to these incidents in order to avoid them in the future. Let us discuss these incidents in detail and the regulations that were implemented following them.

13.2.1 Torrey Canyon

One of the major accidents that resulted in a massive oil spill was the motor tanker *Torrey Canyon*. On 18 March 1967, the 290-metre-long super-tanker registered in Liberia had a cargo capacity of 1,20,000 tons. She ran aground on Pollard's Rock, off the Isles of Scilly, with a full load of crude oil. All efforts to save and refloat the ship failed, and she began to break up. Due to bad weather, efforts to reduce the resulting oil spill were only partly successful. Some of the oil was collected, some was neutralised by spraying dispersants, and some was set ablaze. However, a majority of the oil was dispersed due to the wind and weather. The oil spill resulted in huge environmental damage, including the death of a number of seabirds and other marine organisms. A large amount of oil evaporated, causing air pollution. As a result of this incident, a number of legislations were introduced, the most important of them being the MARPOL Convention, which was adopted in 1973.

13.2.2 Amoco Cadiz

Before the MARPOL Convention could come into force, another major disaster took place. On 16 March 1978, the *M.T. Amoco Cadiz* ran aground on Portsall Rocks, off the coast of Brittany, France, due to the failure of her steering gear. The ship was 334 metres long, and the entire cargo of 220,880 tons of light crude oil was leaked into the sea. The vessel finally broke into three parts in the stormy weather and sank, spilling her entire cargo along with the 4,000 tons of bunker oil that she had on board at the time. The oil spill had a devastating effect on the marine life in the region, causing irreparable damage to the ecosystem. Clean-up operations commenced immediately, but due to the weather and high wind and waves, most of the oil could not be recovered. There is no doubt that a large volume of the oil evaporated, but there has been no specific study or research targeted at this aspect of the oil spill.

In the wake of the *Amoco Cadiz* incident,[72] MARPOL was amended and came to be known as MARPOL 1973 as amended by the Protocol of 1978 (MARPOL 73/78). The protocol introduced new regulations based on the lessons learnt from the spill.

13.2.3 Atlantic Empress

This was the largest oil spill recorded from any ship. On 19 July 1979, two VLCCs (very large crude oil carriers), the *Atlantic Empress* and *Aegean Captain* collided in the Caribbean sea off Tobago66. A sudden rainstorm resulted in bad visibility and heavy weather resulting in the collision. The collision resulted in 287,000 tons of crude oil being spilled from the *Atlantic Empress*. [73] Both the ships caught fire, but firefighters could bring the fire on Aegean Captain under control, and the vessel was towed to Curacao. Unfortunately, 26 crew members from the *Atlantic Empress* and one from the *Aegean Captain* lost their lives. However, the fire on the *Atlantic Empress* continued to rage, and on 3 August the vessel sank, spilling all her cargo into the sea. However, the environmental impact was limited due to the quick response and the nature of cargo, which evaporated over a period of time. But there is no doubt that the oil spill from the *Atlantic Empress* caused considerable air pollution due to the evaporating cargo, as well as the continuous fire and associated massive plumes of smoke.

13.2.4 Exxon Valdez

The oil tanker *Exxon Valdez* ran aground on 24 March 1989, in Prince William Sound, Alaska.[74] This became one of the most famous oil spills and a case study for the shipping industry. The 301-metre-long vessel struck the Bleigh Reef, spilling more than 37,000 tons of crude oil. The size of the spill was relatively smaller than the ones we have discussed so far. However, the waters of Alaska are pristine and ecologically fragile. Consequently, the oil spill caused tremendous damage to the environment, some of which remains to this day. Over 250,000 sea birds and waterfowl were killed, along with sea otters, seals, sea lions, porpoises, and several varieties of birds. Some species have since disappeared from Alaskan waters, such as killer whales and certain varieties of birds like murrelets and pigeon guillemots.

There is no doubt that much of the spilled oil evaporated, but no study was conducted to determine the exact damage to the environment due to the resultant air pollution and release of hydrocarbons. This is commensurate with the global industry's habit of ignoring the intangibles. It is only now that the world has woken to the devastating effects of global warming.

The *Exxon Valdez* incident gave rise to new US legislation known as the Oil Pollution Act 1990, commonly known as OPA 90. The most important requirement of OPA 90 was the Vessel Response Plan. This required both tanker ships and port facilities to have procedures in place to quickly react to oil spills. They should identify a qualified individual in the United States who can respond to oil spills on their ships or ports and coordinate with national agencies.

Moreover, double hull tankers were phased in by the IMO to ensure that even if an incident such as a collision or grounding occurs, only the outer layer of the ballast tank is affected and not the inner cargo tank. This will ensure that pollution does not occur in such cases.

13.2.5 Deepwater Horizon

The oil spill from the oil drilling rig *Deepwater Horizon* is by far the largest ever oil spill in history.[75] On 20 April 2010, the *Deepwater Horizon*, an oil drilling rig, exploded and sank in the Gulf of Mexico, resulting in a catastrophic oil spill and the death of 11 workers, with 17 more injured. The ensuing spill from the pipeline continued for 87 days before the well was finally capped on 15 July 2010. The total amount of oil spilled is estimated to be more than 600,000 tons in addition to 225,000 tons of methane. This unprecedented oil spill caused severe damage to the marine flora and fauna.

The spilled oil was deliberately burnt off to reduce the oil slick. A study by the NOAA and the Cooperative Institute for Research in Environmental Sciences estimated that more than one million pounds of soot (black carbon) was released into the atmosphere, which was equal to the carbon emissions of all ships that travel in the Gulf of Mexico during a nine-week period. In such scenarios, although spills into the sea can be cleaned up and their effects mitigated to some extent, air pollution cannot be avoided. It can only be hoped that nature restores the balance and that such massive spills can be avoided in the future.

In the wake of the incident, British Petroleum, the owner of the rig, was fined to the tune of billions of dollars in claims and penalties. The *Deepwater Horizon* showed a clear lack of safety management on the part of the management. Good safety management, coupled with effective risk analysis, can help avoid such incidents in the future.

Let us now focus on the air pollution that results from an oil spill. Although the impact of oil pollution on the seas is easily visible, the resulting air pollution is invisible to the naked eye and can cause serious damage to the environment.

13.3 OIL SPILLS AND AIR POLLUTION

There have been many serious cases of oil pollution from ships, oil rigs, and shore oil pipelines. Any case of oil pollution invariably leads to the degradation of the environment and air pollution. This is because an oil spill can release airborne pollutants in more ways than one. As discussed earlier in this chapter, evaporation is a major cause of the release of VOC into the air. These suspended gases can also cause acid rain which can cause immense damage to the marine flora and fauna.

The floating oil can lead to the development of aerosols. These are tiny particles suspended in the air. Since they contain oil droplets and other impurities, they can seriously affect human health when inhaled. Marine life such as birds and fishes are also affected by the degraded air quality.

There are other damaging results of oil pollution, and thus immediate action is required to control the oil spill as well as deal with the spilt oil in the water.

However, attempts at pollution control also release dangerous pollutants. Let's see how this can happen:

- Controlled burning of spilled oil is a method often used to mitigate the impacts of the oil spill and prevent it from reaching beaches, fishing grounds, and other inhabited areas. However, the burning of oil releases by-products of combustion into the air, such as soot, particulate matter, sulphur dioxide, carbon dioxide, and NO_x. In large quantities, the harmful effects of these may be even greater than the damage to the seas caused by the spill.
- Chemical dispersants are often used to break down the oil slicks into smaller droplets to increase their natural dispersion. However, these chemicals often release additional pollutants into the air and the seas, thereby causing more harm than good. Only approved dispersants should be used in an oil spill in order to reduce the pollution caused by them.

Consequently, there is a need to analyse the methods used to control oil spills and devise new methods that will not contribute to air pollution.

Let us now discuss a case study related to the efforts of a ship to reduce greenhouse gas (GHG) emissions. It is important to note that this case study is fictional and only for purposes of illustration.

13.4 CASE STUDY: COMPLIANCE AND BEYOND

South Shipping, a fictional shipping company, owns nine bulk carriers. Among these are two handy size vessels of deadweight 28,000 tons, four Panamax vessels of deadweight 70,000 tons, and three cape size vessels of deadweight 220,000 tons.

When the IMO's final strategy for GHG reduction was announced, the company decided to drastically reduce the GHG emissions from their ships by all available means. In early 2023, they requested their classification society to audit their ships to determine their carbon footprint. The results were quite shocking as all their ships were consuming more than the allowed fuel and the CII of their ships ranged between C and D. Realising that this would result in their charterers not renewing the charter parties of their ships, they decided to take whatever measures required to bring the CII rating to A or B.

The leadership recognised the urgent need to implement sustainable practices and ensure compliance with the regulations for GHG emissions. A brainstorming session with the top management, stakeholders, their classification society, and investors was held. Several strategies were discussed and debated before arriving at their final strategy to reduce GHG emissions building up to the introduction of zero-carbon ships by 2040, which would be ten years ahead of schedule.

13.4.1 Strategies Implemented

South Shipping had outlined its ambition to become net zero by 2040. This would give them enough leeway to change their plans in case of any unforeseen circumstances. To this end, they devised two main strategies, the first being to acquire ships operating on alternate fuel and the second being to retrofit some ships to the dual-fuel mode. This would allow the ships to run on alternate fuel and switch back to the conventional fuel in case of supply issues with the alternate fuels. Let's look at these strategies in detail:

- **Alternate Fuels:** It was noted that four of the Panamax vessels will be reaching the age of 20 years in the next three years. A policy decision was taken to scrap or sell these ships before the 20-year special survey (held when the ship reaches 20 years of age) and buy new ships. It was also decided that all new additions to the fleet would be carbon-zero ships. It was known that Maersk Line has already placed the order for green methanol-enabled ships with Hyundai Heavy Industries (HHI), a major shipbuilder in South Korea, the first of which is expected to be delivered in February 2024. It was noted that these ships would be fitted with dual-fuel engines capable of operating on green methanol, as well as biodiesel and conventional bunker fuel. The purchase department was cynical about procuring green methanol, but it was decided to go ahead with this strategy as the production of green methanol would pick up with increasing demand. In this way, the ship can operate on alternate fuel subject to availability and switch back to conventional fuel when required. As the supply of green methanol picks up, it will be increasingly used, and the conventional bunker fuel will be used sparingly, as and when required. It was decided to communicate with HHI to start the formalities of placing the order.

Two of the company's handy size vessels will be reaching 20 years of age in 2028. These ships were about 140 metres long, with a cargo capacity of 25,000 tons. It was noted that in March 2023, Cochin Shipyard in India received an international order and was in the process of building two 135-metre-long container ships powered by ABB's hydrogen fuel system. This was exciting news for the company as the size was similar to their handy size vessels due for the 20-year survey in 2028. The company was sure that by 2028, hydrogen fuel cell technology would have developed, and the cost would be competitive. Further, since Cochin Shipyard was already building such ships it was unanimously agreed to start the formalities for placing an order with them for two handy size vessels powered by ABB hydrogen fuel system.

- **Retrofitting Dual Fuel System:** The three cape size vessels were relatively new, and it was financially not viable to sell or scrap them in order to acquire new carbon-zero ships. These ships were on lucrative long-term charters and the commercial department opined that it would be in the interest of the company to continue with these charters as the market was becoming increasingly competitive. After some discussions and debate, it was decided that retrofitting the engines with a dual-fuel system would be the answer. These ships had engines manufactured by MAN, a major engine maker from Germany. Some quick research revealed that MAN was already into converting big vessels to dual fuel. In fact, they were in the process of converting ships belonging to the container ship management company Seaspan and shipping line company Hapag-Llyod to dual-fuel engines, capable of using methanol and conventional bunker fuel.

  It was decided to start proceedings with MAN for the necessary conversion, starting in 2025. This conversion would be carried out one by one so that only one ship at a time would be out of the market for a prolonged period.

During these brainstorming sessions, the company faced many challenges. Let's discuss how they systematically overcame these challenges.

13.4.2 Challenges Faced

The key challenges faced by South Shipping were as follows:

- The finance department was concerned with the cash outflow associated with these new ideas. They could not understand why a ship that could run for 25 years or more should be scrapped at the age of 20 years. It was explained to them that IMO's stringent GHG strategy did not allow for any laxity and a fall in the CII would result in detentions by port state control and flag state controls, as well as losing market share, since the lucrative charterers would look for ships complying with the new emission standards.

- The commercial department preferred maintaining the status quo as at the time, the company was making a good profit from their ships. The fact that major charterers would look for ships complying with the GHG strategy and having a high CII rating struck a chord with them. Once the commercial aspect was understood, they agreed to go along with the decision to build up to carbon-zero ships by 2040.

Now, let's take a look at some of the key learnings from this case study.

13.4.3 Key Takeaways

In this case study, we have seen how a relatively small company with just nine ships can take proactive steps to comply with the IMO requirements and even go beyond compliance. This ensures that their ships reduce their carbon footprint well before the IMO target dates and become carbon neutral by 2040. The other stakeholders and investors fully supported the idea since the reduction of GHG emissions from ships was high on the global agenda. In addition, they wanted to be seen as promoting the idea of carbon-neutral shipping. Ultimately, it is important to keep in mind that GHG emissions are a real challenge facing the world today. Climate change is real, and all of us in the shipping sector should ensure full cooperation with the IMO strategies so that we can drastically reduce emissions and help address climate change. Groups such as the Sustainable Shipping Initiative (SSI) attempt to drive change through engagement and collaboration with stakeholders to contribute to a more sustainable maritime industry. The SSI had conducted a case study on China Navigation Company regarding ship recycling, AkzoNobel regarding carbon credits, and so on. These case studies[76] show the advantage of sustainable operations on the company's performance and outlook.

Before we conclude, let's look at another case study of how the shipping industry may play a role in increasing environmental pollution during wartime.

13.5 COUNTRIES AT WAR: A CASE STUDY

Warfare results in an exponential increase in environmental pollution due to air strikes, burning of fuel, whole localities going up in flames, deforestation, and even chemical contamination of soil, water, seas and air. This causes long-term damage resulting in loss of biodiversity and can contribute to climate change.

An important case study which establishes the link between war and climate change is the Gulf War, which lasted for several months from August 1990 to February 1991, and reveals the resultant environmental consequences. The deliberate ignition of hundreds of Kuwaiti oil wells by the retreating Iraqi forces resulted in massive fires, lasting for several months. These fires caused irreparable damage to the environment due to the release of massive amounts of black smoke (carbon and soot) as well as greenhouse gases (GHGs) into the atmosphere. This contributed to acid rain, smog, depletion of ozone layer as well as serious adverse effects on human health, agriculture and the ecosystems. In addition, the Gulf War also resulted in environmental damage due to oil spills in the Persian Gulf and destruction of ecosystems due to unregulated military activities.

The Gulf War is a prime example of how armed conflict can result in environmental damage, including the release of GHGs and air and water pollution. Today there are many such hot spots across the world where wars between nations and peoples are damaging the environment. World leaders need to look beyond narrow borders and come together to protect the environment and control climate change.

13.6 SUMMARY

In this chapter, we have seen real-life accidents, as well as a fictional case study. The incidents of the tankers *Torrey Canyon, Amoco Cadiz*, and *Exxon Valdez* were devastating for the marine environment. It is important to study these cases to build awareness of preventive actions, in the form of regulations, guidelines, circulars, and so on, that can be taken in such scenarios.

IMO was quick to learn from these incidents and brought in regulations to prevent re-occurrence. The case study of the fictional shipping company, South Shipping, serves as an eye opener to companies as to how they quickly come to terms with the IMO GHG strategies and be proactive in compliance well before the target dates.

In the final chapter of this book, we shall discuss the future regulations and initiatives by the IMO in particular and the shipping industry in general to ensure that the 2023 IMO GHG strategy is successful in every way.

BIBLIOGRAPHY

71. ITOPF Technical Paper on Marine Oil Spills. https://www.uvm.edu/seagrant/sites/default/files/uploads/TIP2FateofMarineOilSpills.pdf.
72. Story of the *Amoco Cadiz* Oil Spill. https://www.resources.org/archives/an-oil-spill-and-its-aftermaththe-story-of-the-amoco-cadiz/.
73. Stuart A. Horn and Captain Phillip Neal. *The Atlantic Empress Sinking*. https://shipwrecklog.com/log/wp-content/uploads/2014/10/mobil-atlanticempress.pdf.
    This report by Mobil Oil Corp details the collision which resulted in the catastrophic oil spill from the *Atlantic Empress*.
74. Article on *Exxon Valdez* Oil Spill. (2018). https://corporate.exxonmobil.com/who-we-are/technology-and-collaborations/energy-technologies/risk-management-and-safety/the-valdez-oil-spill#ChangesExxonMobilhasmadetopreventanotheraccidentlikeValdez.
    This article by ExxonMobil gives the company's perspective of the oil spill from their vessel, *Exxon Valdez*.

75. Report on the Deepwater Oil Spill. (August 2023). https://www.epa.gov/enforcement/deepwater-horizon-bp-gulf-mexico-oil-spill.

This report by the United States Environmental Protection agency gives the current picture of the incident as well as the court proceedings to date.

76. Some Case Studies by Sustainable Shipping Initiative. https://www.sustainable-shipping.org/category/resources/case-studies/.

Few case studies are discussed in order to underline the importance of sustainable operations.

Chapter 14

The Way Forward

So far, we have discussed the various methods that ships and shipping companies can adopt to reduce greenhouse gas (GHG) emissions and control global warming. In this chapter, we will have a look at what the future holds for us. We shall discuss the various Marine Environment Protection Committee (MEPC) resolutions and understand what they mean. But first, what is the MEPC? International Maritime Organization (IMO) constituted the MEPC to look after various aspects related to the protection of the environment. The MEPC consists of the representatives of the member states of IMO. They are responsible for the development and implementation of international regulations aimed at reducing pollution from ships and protecting the marine environment. They hold regular sessions to address various environmental issues related to the shipping industry. During these sessions, they issue resolutions, guidelines, circulars, or amendments based on the discussions during the session.

MEPC resolutions form the backbone of IMO's efforts to protect the environment. Compliance with these resolutions will ensure that the ship causes minimum damage to the environment during her operations. IMO has stepped up efforts to prevent global warming. MEPC 80 is the landmark resolution related to air pollution and GHG emissions. MEPC 81, 82, 83, 84, and so on all spell out the future efforts required to prevent the escalation of global warming and ensure environmental sustainability in the shipping sector. It is important for us to understand what the future holds, and hence, an understanding of these resolutions is required for both shore and ship staff in order to put these resolutions into action. Through these resolutions, the IMO is building the road map to zero-carbon shipping by 2050 and leading the way to the sustainable development of the shipping sector.

Let us now have a look at the comprehensive impact assessment (CIA) initiated by MEPC 80 in order to assess the impact of the GHG reduction measures on the stakeholders.

DOI: 10.1201/9781032702568-14

14.1 INTRODUCING THE COMPREHENSIVE IMPACT ASSESSMENT

The CIA is an evaluation process mandated by the IMO in MEPC 80. It is an examination of the effects of the measures introduced by IMO on member states, ship-owners, and other stakeholders.[77] These measures include the EEDI, energy efficiency existing ship index (EEXI), CII, and so on.

The CIA should be overseen by a steering committee set up by the IMO, comprising representatives of member states and coordinated by the vice-chair of the MEPC. The steering committee is expected to table the CIA report as well as recommendations during MEPC 81, so that the committee can make decisions based on these facts and figures. These recommendations will help the IMO to make course corrections to the GHG reduction measures and to introduce new measures if required. It must be in alignment with both maritime activity and business-as-usual (BAU) emission scenarios. BAU is the effect on the environment if we continue with the current trend without making any extra efforts to curb GHG emissions. It should also ensure the coherence of emission reduction scenarios across various situations, in adherence to the 2023 IMO GHG Strategy. This process forms the foundation for the final selection of the combination of measures by MEPC 81 in March 2024. Thus, the CIA is not about pure compliance with the regulations. It is more holistic in that it also takes into account the practical realities of the shipping industry and integrates this with the IMO strategies. This will help motivate the ship-owners to adhere to these requirements without endangering their commercial commitments. The CIA will consider the following factors:

- Geographic remoteness of and connectivity to main markets
- Cargo value and type
- Transport dependency
- Transport costs
- Food security
- Disaster response
- Cost-effectiveness
- Socio-economic progress and development

The CIA should comprised the following elements:

- **Literature Review:** The report reviews existing research and knowledge on the subject, providing context and insights into the findings. The methodology and scope of the assessment study are mentioned in the review.
- **Assessment of Impacts on the Fleet:** This gives the details of how the proposed measure will affect the fleets involved, such as operational changes, costs, or efficiency.

- **Assessment of Impacts on States:** This element specifies how the measure will impact different states involved. Economic consequences, regulatory implications, social, and environmental impacts are some of the parameters measured.
- **Stakeholders' Analysis:** Analysing the perspectives and concerns of stakeholders are an important aspect of the CIA in order to understand their viewpoints and repercussions of the implementation, both positive and negative.
- **Identification of Missing Data:** Acknowledging limitations in data, ensuring data quality, and conducting uncertainty and sensitivity analyses are necessary for the assessment to give a true picture of the impact.

We have seen the efforts of IMO aimed at reducing GHG emissions. In order to ensure net-zero emissions by 2050, IMO has lined up a series of further MEPC sessions to ensure the momentum gained by the previous resolutions and actions required thereof. Let us discuss these future resolutions planned by IMO.

14.2 FUTURE RESOLUTIONS

In addition to MEPC 80 and 81, there are several sessions planned with the aim of streamlining the efforts to reduce GHG emissions from ships and promote sustainable development. The future IMO resolutions and their requirements can be classified as follows:

- MEPC 82: The final report of the CIA is tabled and considered by the committee for bringing about amendments to MARPOL Annex VI based on these findings.
  Prior to this, a two-day expert workshop (Fifth GHG Expert Workshop – GHG-EW 5) will be held to consider the CIA (on member states) and report the outcome to MEPC 82.
  The 17th ISWG on GHG emissions (ISWG-GHG 17) to consider the outcome of CIA, the GHG-EW5, and report to MEPC 82. ISWG-GHG 17 will also develop the terms of reference of a Fifth GHG Study.
- MEPC 83: Approval of the amendments to MARPOL Annex VI and review of the short-term measures which are to be completed by 1 January 2026.
- Approval of amendments to MARPOL Annex VI based on the findings of the CIA
- MEPC 84: Review of the short-term measures
- MEPC 85, 86, and 87 are still not finalised but will be a review of the various aspects of reduction of GHG emissions.

- MEPC 88: This will include a review of the earlier resolutions which will finally lead to the 2028 strategy on GHG.
- Autumn 2025: An extraordinary MEPC session for adopting the MARPOL amendments relating to GHG reductions. These amendments are not yet decided as they will depend on the findings of the CIA.
- Sometime in 2027: Entry into force of the above amendments (basket of measures) 16 months after the adoption.

To put the whole concept of CIA and future regulations in perspective, let us have a close look at both the technical and the economic element to be considered.

The ISWG regularly meets between the MEPC sessions to firm up details for discussion and adoption at the MEPC sessions. As per IMO Resolution MEPC 80, the candidate mid-term measures for reduction of GHG emissions include the technical element and the economic element. Let us have a look at these and how they will develop in future so that the IMO 2023 GHG strategy can be achieved:

- **Technical Element:** A goal-based marine fuel standard regulating the phased reduction of the marine fuel's GHG intensity.
  An example of this is establishing a greenhouse gas fuel standard (GFS). With a predictable increase of ships operating on alternate fuels in the future, global availability of alternate fuels and technologies is to be ensured. The GFS will enable production of these fuels. With demand picking up for low and zero GHG fuels, the bunkering infrastructure should also pick up. The greenhouse gas fuel intensity (GFI) will be gradually reduced over a period of time, thereby allowing time for the fuel transition and minimising their impact on the economy. The GFI reduction pathway is to be developed for approval by MEPC 83.
- **Economic Element:** An economic element on the basis of a maritime GHG emissions pricing mechanism.

For instance, the introduction of the flexibility compliance mechanism (FCM) grants flexible compliance units (FCUs) to ships that go beyond required compliances of GHG emissions. The GFS along with the FCM will be enforced to ensure consistency with UNFCCC and Paris Agreement requirements in order that global warming in controlled.

So far, we have seen the requirements of MEPC 80 and MEPC 81, including the CIA. The main aim of these resolutions is to reduce GHG emissions within the shipping industry. IMO is also coming up with a slew of resolutions within the next few years to ensure that the 2050 ambition of net-zero shipping will be achieved on time. In this section, we will discuss these IMO resolutions. Details are sketchy, as they are still in the preliminary stages, but we can get an idea of what is coming next in the field of energy efficiency and sustainable development in shipping.

14.2.1 IMO's Future Planning

IMO is working to ensure that the GHG strategy is successful and shipping can achieve net-zero within the target date. The immediate target of IMO is to ensure that the mid-term GHG reduction measures result in a smooth transition of shipping with a level playing field and a just and equitable transition to zero emissions.[78]

In order to provide the world's fleet an incentive towards this, a pricing mechanism is important. The IMO 2023 strategy includes a time line for the adoption of these measures:

- Approval of mid-term measures at MEPC 83 during the spring of 2025.
- Adoption of mid-term measures at an extraordinary session of MEPC. In the extraordinary session, the measures will involve the IMO developing a basket of mid-term measures, leading to reduction targets.
- Entry into force of mid-term measures in 2027, that is, 16 months after adoption.

These mid-term measures include the technical measures and the economic measures, as explained earlier.

That brings us to the next IMO resolution MEPC 82, which will consider the final report of the CIA.

14.2.2 MEPC 82

In this session of the MEPC which is scheduled for 30 September to 4 October 2024, the final report of the CIA will be tabled and considered.

Based on the final combination of measures recommended by MEPC 81, a further CIA will be conducted by the steering committee who will then table the final report of the CIA at the MEPC 82.

14.2.3 MEPC 83

In this session of the MEPC to be held sometime during the spring of 2025, approval of amendments to MARPOL Annex VI based on the findings discussed in MEPC 82 will be accorded. It includes the approval of the mid-term measures and the review of the short-term measures which is to be completed by 1 January 2026.

14.2.4 MEPC 84

This session is scheduled to be held in the spring of 2026 with the aim of approving and reviewing the short-term measures (EEXI and CII) that are scheduled to be completed by 1 January 2026.

14.2.5 MEPC 85

The 85th session of the IMO's Marine Environment Protection Committee (MEPC): This session has not yet occurred and is tentatively scheduled for autumn 2026. As of today, 19 February 2024, the specific agenda and outcomes of this session are unknown. However, based on the ongoing focus on reducing GHG emissions from shipping, it's likely that MEPC 85 will address related topics.

14.2.6 Other Resolutions

The following MEPC sessions are scheduled in future to deal with the various aspects of reduction of GHG emissions. However, the exact agenda is not yet finalised and will depend on the outcome of the previous sessions:

- **MEPC 86:** This session is scheduled to be held in the summer of 2027 – review of the 2023 IMO strategy.
- **MEPC 87:** This session is scheduled to be held in the spring of 2028.
- **MEPC 88:** This session is scheduled to be held in the autumn of 2028 – Finalisation of the review of the 2023 IMO GHG strategy leading to the adoption of the 2028 IMO GHG strategy.

IMO has included a wide variety of design, operational, and economic solutions to achieve the target of net-zero emissions by 2050. Their GHG reduction potential is estimated as follows:[78]

Voyage optimisation: 1–10%
Energy Management: 1–10%
Hull and superstructure: 2–20%
Concept, speed, and capability: 2–50%
Power and propulsion systems: 5–15%
Hull biofouling management: 2–20%
Fleet management, logistics, and incentives: 5–50%
Bio-LNG/LPG: 35%
Full electric: 50–90%
Extensive speed optimisation: Up to 75%
Hydrogen and other synthetic fuels: 80–100%
Biofuels third generation: 90%

It is left to the ship-owners and operators to select the method of GHG reduction so that their ships meet the target GHG emission levels in time.

14.3 TAKING STOCK OF THE FUTURE

It appears that the shipping industry is on track to meet the target of IMO to make shipping zero-carbon by 2050. In general, we can state the future developments of GHG reduction as follows:

- Methanol is one of the best alternatives to fossil fuel. It has lower risk of flammability and is cost-effective. Thus, methanol engines are a good option for decarbonisation, and many companies are ordering ships that can use methanol as fuel.
- Green hydrogen does not emit any polluting gases and is thus is a sustainable alternative to fossil fuel. But it is not cost-effective, and there are safety issues in the use of hydrogen. Once these are sorted out with advanced technologies, green hydrogen would be the best option for zero-carbon ships.
- Carbon capture on board is beneficial because ships can continue using fossil fuel without GHG emissions but is expensive, and infrastructure needs to be ramped up. Carbon capture can be considered to be a short-term measure for existing ships while the company transitions to new ships that can operate on alternate fuels or renewable energy.
- Dual-fuel engines can be retrofitted on existing ships and is an attractive option. Ships that have diesel engines can convert the single-fuel engines to dual-fuel and thereby reduce their carbon footprint.
- New ships fitted with dual fuel engines and energy saving technologies are becoming more popular.

Many ship-owners are going beyond compliance and trying to achieve the IMO 2050 targets well before time. But for this, ship-owners will have to incur additional expenses, such as for retro-fitting dual-fuel engines, ordering the more expensive ships operating on alternate fuel, renewable energy, and so on. This is the way forward for the maritime sector in order to ensure sustainable development.

14.3.1 Corporate Social Responsibility and Environmental, Social, and Governance

Corporate social responsibility (CSR) is a management principle in which companies incorporate social and environmental considerations into their business practices and engagements with stakeholders. It strikes a balance between economic, environmental, and social goals (known as the "Triple-Bottom-Line Approach", while meeting the expectations of shareholders and stakeholders. Companies incorporate CSR in their vision statement, thereby making their services and products acceptable to their clients and the public at large. A company practising CSR will incorporate social, environmental, and ethical considerations into the company's operations. In short, the goal of CSR is to encourage businesses to be socially accountable for their actions, while ensuring the long-term sustainability of their operations. CSR has been around for a long time but has become more popular in corporate circles during the last 50 years or so.

Environmental, social, and governance (ESG) is a more recent concept, which is increasingly becoming popular. The maritime industry has embraced the concept of ESG with vigour. A company which has obtained a certificate of ESG compliance will have a competitive advantage over their peers. Certification entities such as classification societies will analyse all aspects of the shipping company's operations to ensure that the company adheres to the highest standards of ESG.

Factors such as GHG emissions, recycling, ecological impacts, health and safety, accident and safety management as well as business ethics are the parameters to be reported for ESG certification for shipping companies.

ESG is considered as an advancement over CSR as it is concerned more with the aspects of environmental sustainability.

14.4 SUMMARY

This chapter explored the forthcoming IMO resolutions poised to shape the future of energy efficiency in shipping. These resolutions study the impact of the existing IMO regulations and thereby are expected to chart the future course of action. This is keeping in mind the impact of such requirements on the stakeholders in order to achieve sustainable development. This underscores the functioning of IMO, where the needs and limitations of the member states are taken into consideration while deciding on new regulations to reduce GHG emissions from ships. It is important to be aware of these resolutions because understanding them will help us to comply with international regulations, promote sustainable development of the shipping industry, and participate in global efforts to protect the marine environment.

We are now at the end of this book, which was a comprehensive exploration of the crucial relationship between maritime transportation and sustainable practices. We have delved into the complex subject of energy efficiency within the shipping industry and its implications for environmental protection and sustainable development. We have discussed the importance of preventing global warming by reducing GHG emissions, which is possible by energy-efficient operations on board ships, including reducing the dependence on fossil fuels. We reiterate the importance of using advanced technology, alternate fuels, and renewable energy.

The IMO is continuously attempting to reduce GHG emissions from ships by means of amendments to Annex VI of MARPOL and various MEPC resolutions. These are discussed in detail throughout this book in an attempt to focus on the regulatory requirements. Compliance with these requirements is necessary for the ship to trade across continents. More importantly, they are required to ensure that GHG emissions are reduced and climate change is prevented.

It is hoped that this book will serve as a valuable resource for both students and stakeholders in the shipping industry to deepen their understanding of energy efficiency in shipping and contribute to the advancement of sustainable development in the maritime industry.

BIBLIOGRAPHY

77. Assessing CIA Impacts. https://wwwcdn.imo.org/localresources/en/OurWork/Environment/Documents/MEPC.1-Circ.885-Rev.1.pdf.
78. IMO's Work to Cut GHG Emissions from Ships. https://www.imo.org/en/MediaCentre/HotTopics/Pages/Cutting-GHG-emissions.aspx.

Index

For Product Safety Concerns and Information please contact our EU
representative GPSR@taylorandfrancis.com
Taylor & Francis Verlag GmbH, Kaufingerstraße 24, 80331 München, Germany